Biocultural Histories in *La Florida*

A Dan Josselyn Memorial Publication

Biocultural Histories in *La Florida*

A Bioarchaeological Perspective

Christopher M. Stojanowski

THE UNIVERSITY OF ALABAMA PRESS
Tuscaloosa

The University of Alabama Press
Tuscaloosa, Alabama 35487-0380

Manufactured in the United States of America

Typeface: Minion

∞
The paper on which this book is printed meets the minimum requirements of American National Standard for Information Sciences-Permanence of Paper for Printed Library Materials, ANSI Z39.48-1984.

Library of Congress Cataloging-in-Publication Data

Stojanowski, Christopher M. (Christopher Michael), 1973–
Biocultural histories in La Florida : a bioarchaeological perspective / Christopher M. Stojanowski.
p. cm.
Includes bibliographical references and index.
ISBN-13: 978-0-8173-1485-9 (hardcover : alk. paper)
ISBN-10: 0-8173-1485-7 (hardcover)
ISBN-13: 978-0-8173-5267-7 (pbk. : alk. paper)
ISBN-10: 0-8173-5267-8 (pbk.)
1. Indians of North America—Missions—Southern States. 2. Indians of North America—Anthropometry—Southern States. 3. Indians of North America—Southern States—Population. 4. Missions, Spanish—Southern States—History. 5. Ethnoarchaeology—Southern States. 6. Human remains (Archaeology)—Southern States. 7. Dental anthropology—Southern States. 8. Florida—History—Spanish colony, 1565–1763. 9. Southern States—Antiquities. I. Title.
E78.S65S75 2005
304.6′09759′09032—dc22

2005013631

Contents

List of Illustrations

Figures

Tables

Acknowledgments

The list of acknowledgements for this project is extensive. The following individuals accommodated me during the data collection phase of this project: Clark Spencer Larsen, Jerald Milanich, Dale Hutchinson, Bram Tucker, Christopher Rodning, David Hally, James Krakker, and David Hunt. I would also like to thank my dissertation committee members, Jane Buikstra, Joseph Powell, Lyle Konigsberg, and Edward Bedrick, for providing helpful insight into this project. Conversations with Bonnie McEwan and Rochelle Marrinan also greatly improved my personal understanding of the mission period in Florida. Rochelle's 16-week field school at the O'Connell Mission site initiated my interest in this period of Florida's history. Prudence Rice and the Office of Research Development and Administration at SIU-Carbondale graciously approved research assistant funding without which certain deadlines would have passed unmet. I would like to acknowledge William Duncan for agreeing to work as said assistant. Finally, my family deserves special recognition for their patience and support during the final writing phases of this project. Grant funding for data collection at the Smithsonian Institution, the University of Georgia, the University of North Carolina–Chapel Hill, and the Florida Museum of Natural History was provided by the Wenner-Gren Foundation for Anthropological Research (grant no. GR-6698), the University of New Mexico Graduate School, and Sigma-Xi.

Biocultural Histories in *La Florida*

1 Historical Bioarchaeology

In 1763, Spanish ships set sail from the Atlantic Coast of Florida heading toward Havana, thus ending two centuries of Spanish colonization and proselytizing of the region. These ships carried the few remaining individuals from the indigenous colonial populations (the Apalachee, Guale, and Timucua) that had survived decades of social upheaval and alterations in traditional lifeways. Although Spain would return to *La Florida* twenty years later for a brief period of occupation, the Native American populations did not. The ceding of St. Augustine in 1763 was, therefore, the culmination of a long process of Native American acculturation, resistance, and adaptation initiated by Ponce de León's 1513 Gulf Coast landing and subsequent claiming of lands north of New Spain for the Spanish Crown. Historical documents provide considerable insight into what life was like during the contact period. They are not, however, complete. And, for the biological anthropologist, the evolutionary consequences of the mission experience remain unclear. This book investigates patterns of biological variability in light of the mission history to supplement the historical record in a way that will be useful to historians, archaeologists, and physical anthropologists.

Specifically, this book investigates changes in population levels of phenotypic variability for two tribal groups (the Guale and Apalachee) through three time periods, Late Precontact (~A.D. 1200–1400), Early Mission (A.D. 1600–1650), and Late Mission (~A.D. 1650–1706). By focusing on comparisons within groups, I evaluate the pattern of changes in phenotypic (tooth size) variability (increase, decrease, or stasis) and relate observed trends to prevailing models of New World population demography and ethnohistoric details of population structure and interaction patterns. Ultimately I hope to reconstruct the evolutionary impact of Spanish policy and Native American response to this policy, and further detail the rapidly changing sociopolitical world in which the indigenous populations of *La Florida* found themselves.

The social history of Spanish Florida identifies two overarching processes defining changes in population composition through time. First, population sizes were declining, albeit at different rates in different regions. Second, population aggregation and migration were occurring largely in response to the rate of localized demographic collapse. Aggregation can therefore be viewed as secondary to population demography and as a reactionary process designed to mitigate the deleterious effects of demographic collapse. Where demographic collapse was most severe, the hierarchical process of population aggregation was most aggressive. In this sense, "hierarchical" refers to the fact that aggregation proceeded along a defined, progressive pathway, from the local to the regional to the supraregional. Expansion of the "burial catchment," that is, the geographical area from which a particular mission community cemetery received deaths, has several predicted effects depending on the structure of population relationships in preceding time periods. Aggregation of biologically integrated populations is an evolutionary non-event, whereas aggregation of divergent populations (and admixture between them) leads to predictable evolutionary genetic responses.

Ethnohistory's contributions to this discussion are twofold. First, ethnohistoric data provide estimates of population size that are directly related to expected intensities of population aggregation. Quality varies considerably (see Henige 1998), a topic returned to throughout this book. Second, ethnohistory provides statements regarding the expected biological consequences of aggregation, where it did occur. In other words, ethnohistoric data can be used to define population interaction boundaries, to predict changes in the size of the population, and to predict diachronic changes in between-group interaction patterns.

This book considers the effects of "collapse aggregation" in terms of synchronic and diachronic differentials in genetic variability. Where there is documented stasis or a decrease in genetic variability, paleogenetic bioarchaeology informs about rates of population size reduction and the effects of genetic drift. For example, a decline in genetic variability suggests a similar decline in population size. Static genetic variances suggest no change in population size between time periods. Where there is a documented increase in genetic variability, paleogenetic bioarchaeology informs about precontact population structure and patterns of gene flow during the contact period. For example, knowledge of contact between two populations during the historic period, combined with genetic variability estimates for both populations during sequential time periods, allows inference about the degree of biological integration of these populations during the earlier time period. If variability increases, it is assumed the populations were previously genetically distinct. Assuming veracity of the contact era data (population size debates notwithstanding), the approach adopted in this book has the ability to evaluate simultaneously the accuracy of archaeologically and historically generated models of biosocial interaction both before and during the period of active missionary activity. It also provides novel information about the nature of such interaction unavailable from other sources. One facet of this book is, therefore, expository; to demonstrate the utility of paleogenetic bioarchaeology in anthropological research programs.

Attempts to tether diverse topics into a coherent picture of the social environment of 17th-century *La Florida* necessitate a careful distillation of data and presentation of research foci. The success of this approach requires delineation of three topics: (1) general historical processes must be outlined and used to define a research design; (2) analytical data must be explained and explicitly linked to the research design; and (3) analytical results based on the data must be evaluated in light of the research questions and regional histories.

With this structure in mind, I have organized the book into three sections. The first, "The Archaeology," consists of chapters 2 and 3. Chapter 2 is written to provide general historical and archaeological details and to contextualize the current work within the appropriate regional scholarship. We know the dates, personalities, and places of *La Florida.* We know the effects of missionization on native health and diet. We know epidemics were problematic. We know social systems were stressed. We do *not* know the evolutionary impacts of colonization. Whereas chapter 2 provides a general historical overview, chapter 3 specifically targets data sources integral for generating biologically informed predictive models. It is in this chapter that disease epidemiology is incorporated into the research design and that Spanish policies to mitigate ensuing colonial stresses are outlined. This discussion is structured around the central overarching theme of this book, namely the complementary nature of ethnohistoric and bioanthropological data.

The second section, "The Bioanthropology," consists of chapters 4 and 5. Chapter 4 links the ethnohistory with evolutionary theory, with discussion centered on the causes and interpretation of phenotypic variability, the meaning of "heritability," odontometric inheritance patterns, and evolutionary mechanisms effecting variance transitions. Chapter 5 is methodological in focus and continues discussions of odontometric research. Statistical analyses and pretreatment protocols are presented.

The final section, "The Synthesis," presents results and interpretations of the analyses. Chapter 6 discusses phenotypic transformation within the context of Apalachee archaeology and history; chapter 7 discusses similar results for Guale. Both chapters present discussion and interpretation within regional ethnohistories. Chapter 8 presents comparative and concluding remarks and considers this project in a broader, regional perspective.

Broader Impacts

This book is overly broad in academic scope. However, the question of interest is rather straightforward. What effects did colonization have on indigenous populations, and how did the response to prevailing biosocial conditions manifest in population genetic patterns of variability and affinity? Specifically, the following questions are addressed:

(1) What pattern of biological interaction characterized the precontact period?
(2) Was the pattern consistent with etic or emic ethnic divisions at the time of contact?
(3) How did this pattern change through time?
(4) What may have caused observed changes or explain differential patterns of change between different provinces?
(5) How did population sizes change during the mission period?
(6) Was change consistent with historical predictions?
(7) Was change consistent between provinces?

The larger issues developed in this book, however, are of broader interest. Bioarchaeology is well positioned within anthropology to unify evolutionary and cultural concepts under both historical and processual perspectives. As alluded to earlier in this chapter, and as demonstrated in the following discussion, bioanthropological and ethnohistorical data are complementary information sources. Neither is free of underlying and unverifiable assumptions. This case study demonstrates how the interplay of archaeology, history, ethnohistory, linguistics, and paleogenetic bioarchaeology generates a better informed model of anthropological inference.

Part I
The Archaeology

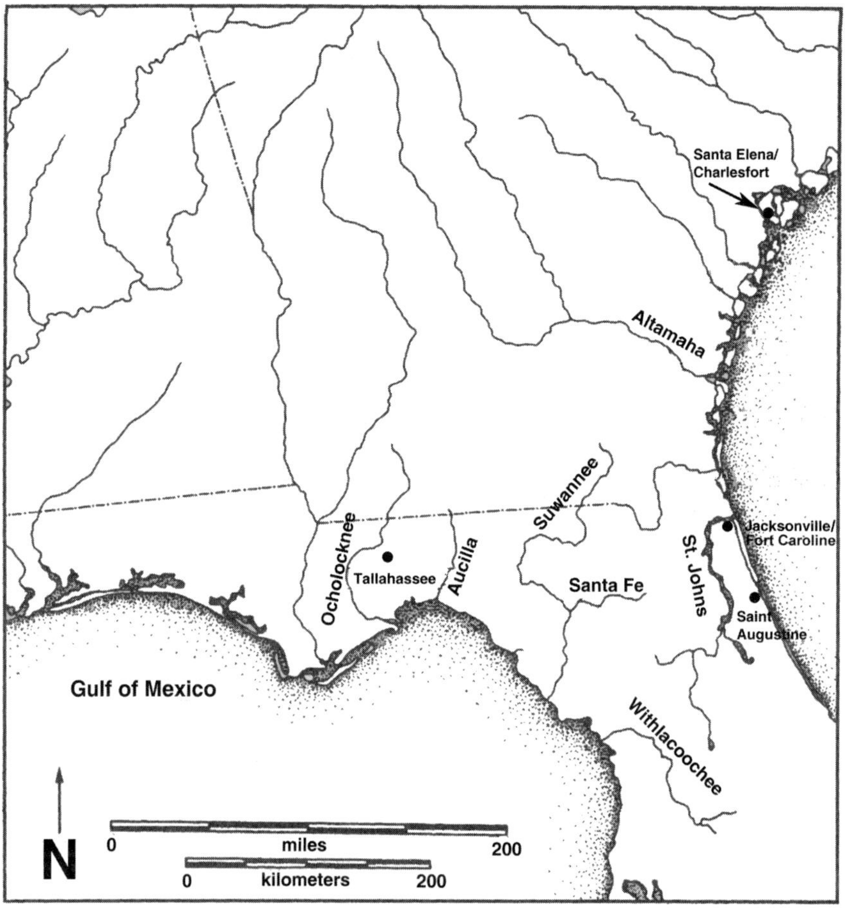

Figure 2.1. Map of important locations and rivers in *La Florida* (modified from Larsen 2001: Fig. 2.1).

2 The Setting

The Spanish Mission System of *La Florida*

> One of Menéndez's captains thrust his dagger into Ribaut's bowels, and Merás, the *adelantado's* brother-in-law, drove his pike through his breast; then they hacked off his head. Some there were . . . who condemned Menéndez for his cruelty and for slaying the captives after having given his oath for their safety. But Barrientos . . . holds that he was "very merciful" to them for he could "legally have burnt them alive. . . . He killed them, I think, rather by divine inspiration."
>
> (Bolton 1921:148–149)

Founded by Ponce de León in 1513 and colonized by Pedro Menéndez de Avilés in 1565, Spain's *La Florida* colony (Figure 2.1) represented the beginning of permanent European colonization of North America, a process that would ultimately result in widespread ethnocide of myriad indigenous cultures. In its most general sense, this book considers the processes responsible for the extinction of these communities and the active strategies adopted to counteract impending demographic catastrophe.

Table 2.1 summarizes key historical details for those unfamiliar with this regional context. In-depth historical discussion of Spain's North American presence can be found in multiple sources (Bolton 1917; Boyd 1948; Gannon 1983; Geiger 1937; Griffin 1990; Hann 1986a, 1988, 1993, 1996; Lyon 1990; McEwan 1993; Milanich and Proctor 1978; Smith and Gottlob 1978; Spellman 1965; Sturtevant 1962; Thomas 1990a), but I will review major events here.

The history of Spain's involvement in the New World is complex and has been the subject of intensive anthropological and archaeological investigations for nearly a century. The province of *La Florida* was named in 1513 when Juan Ponce de León landed near the Tampa Bay/Charlotte Harbor area and claimed all lands east and north of New Spain for the Spanish Crown. He named the colony after the *Pascua Florida*, the Feast of Flowers, which preceded the Easter holiday and commemorated the date of his landing. Because Spain already possessed extensive holdings in South and Central America and the Caribbean, the colonization of *La Florida* was deemed necessary for three reasons: to create a buffer zone against French colonial interests farther north (Lyon 1990; Milanich 1990; Sluiter 1985; Spellman 1965); to protect shipping lanes for the transportation of gold from New World colonies to Europe (Sluiter 1985;

Table 2.1. Important historical dates for the First Spanish Period

Date	Event
1513	Juan Ponce de León claims *La Florida* for Spain
1528	Narváez entrada
1539–1540	De Soto entrada
1565	St. Augustine founded
1566	Jesuits arrive and begin widespread effort at conversion throughout colony
1572	Jesuits leave Florida
1573	Franciscans arrive
1597	Guale revolt
1607	Franciscans return to Guale
1607	Western Timucua receives missionaries
1633	Apalachee receives missionaries
1661	Westos begin attacks on Guale
1684	Guale abandoned
1685	Carolina colony, under Moore, begins assaults on Apalachee and Timucua provinces
1704	Moore raids Apalachee, province is abandoned
1706	Moore raids Timucua, province is abandoned
1763	Spain cedes its North American holdings (St. Augustine)

Weddle 1985); and to establish a terrestrial route from the Atlantic Ocean to New Spain (Milanich 1990; Pearson 1968).

Initially, Spanish *entradas* (expeditions) were staged in the hope of recovering wealth in the form of precious metals or slaves to replenish *encomiendas* (a labor land grant system) in the Caribbean, where demographic collapse had been nearly complete (Hutchinson 1990; Lyon 1991). After several half-hearted attempts at colonization, the French established Charlesfort and Fort Caroline on the southeastern Atlantic coast (Figure 2.1), which hastened Spanish interest in permanently colonizing North America (Lyon 1991; Thomas 1990a). St. Augustine, Spain's answer to the French threat, was established in 1565 by Pedro Menéndez de Avilés (Barrientos 1567; Lyon 1991) and commonly carries the appellation of "America's oldest city."

With Menéndez's capital city of St. Augustine strategically situated on the northern Atlantic coast of Florida (Figure 2.1), Spain's Catholic ambassadors immediately sought redemption for the pagan indigenous groups as part of the *conquista de almas* (conquest of souls). In fulfillment of his royal charter, Menéndez at first enlisted Jesuit missionaries (1566–72), followed by Francis-

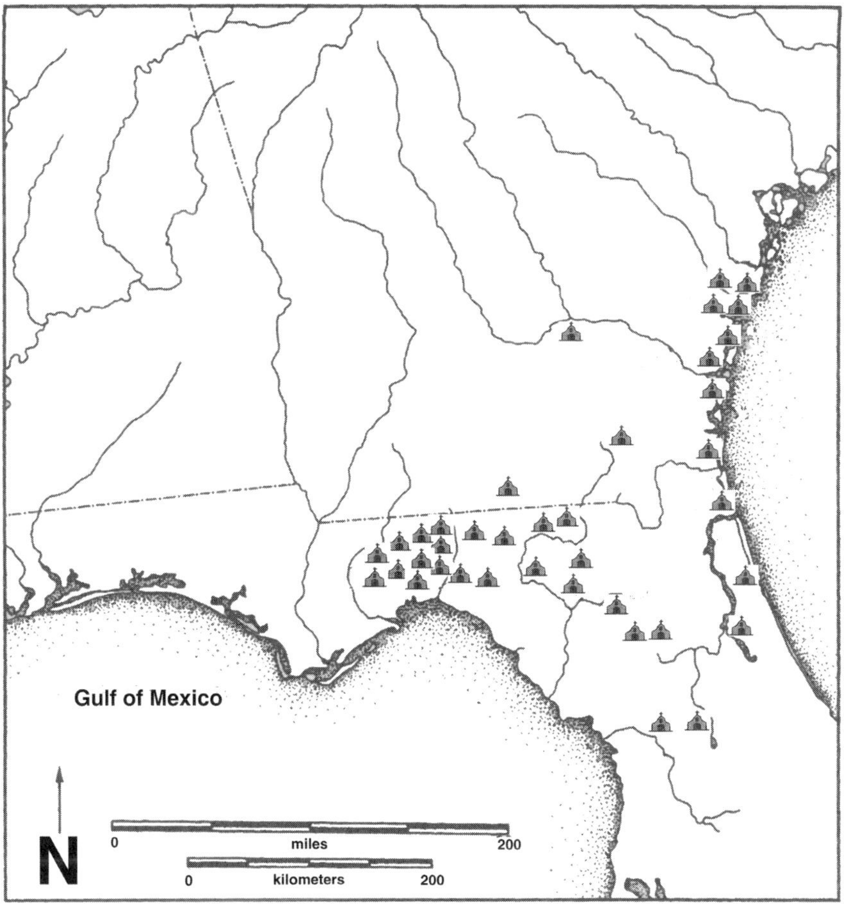

Figure 2.2. Map of mission locations ca. 1650 (modified from Larsen 2001: Fig. 2.1).

cans (1573–1706), to organize the conversion and pacification of the indigenous populations (Deagan 1985; Geiger 1937; Hann 1991; Milanich 1990; Thomas 1990a; Weisman 1992). Initially, mission efforts were restricted to the area immediately surrounding St. Augustine (Sturtevant 1962) (Figure 2.2). Early success, however, quickly led to the expansion of missionaries into southern coastal Georgia and South Carolina, the domain of the Guale, Escamaçu, and Orista (Jones 1978; McEwan 2001), followed by the St. Johns River drainage (eastern Timucuan speakers) and northern Florida interior (western Timucuan groups) (Deagan 1990a; Smith and Gottlob 1978). The Apalachee chiefdom (roughly encompassing Leon and Jefferson counties near present-day Tallahassee) received Christianity in 1633 (Hann 1988), following a 25-year period of intermittent missionary contact. By 1650 the *La Florida* mission chain extended north along the Georgia coast and west through the provinces of Timucua and Apalachee (Figure 2.2).

Taking 1633 as the date of maximal mission expansion (and ignoring late-17th-century movement beyond Apalachee into the Apalachicola region; see Hann 1988), the "Golden Age" of the mission period can be defined as the period from 1633 until approximately 1661 (Geiger 1937; Spellman 1965). In this year, the mission provinces, beleaguered by nearly a century of political unrest, native revolts, and disease epidemics, came under attack by English-allied indigenous populations. Initial conflict centered on the Guale chiefdom of coastal Georgia, where, beginning in 1661, slave-raiding attacks by the Chichimecos, Creeks, and Cherokees began (Covington 1968; Worth 1995). Conflict would ultimately cause the abandonment of Guale by 1684 (Worth 1995) and a resulting southward contraction of pro-Spanish populations toward the stronghold of St. Augustine. Continued development of English interests in the Carolinas and Virginia (in particular the establishment of Charlestown in 1670) would ultimately spell the end of the mission system. Beginning in 1685, English-allied Indians began attacking the provinces of Apalachee and Timucua (Hann 1988; Thomas 1990a). By 1704, many missions in Apalachee had been burned and the population scattered (Boyd et al. 1951; Hann 1988). Some natives were taken prisoner by the English and returned to the Carolina colony; others fled to Mobile to seek French protection, and a minority, remaining Spanish loyalists, fled to St. Augustine (Covington 1964; Hann 1988). The western Timucuan missions were similarly destroyed in 1706 (Hann 1996; Milanich 1996), and, as with the Apalachee, the survivors either escaped to the woods or fled to St. Augustine where populations continued to dwindle throughout the 18th century until Spain abandoned Florida in 1763. This retreat to Havana marks the official end of the First Spanish Period in North America, and with it ended the history of northern Florida's and southern Georgia's indigenous populations.

The Mission Populations

Modern scholars typically divide the mission populations into three provinces (Figure 2.3) that corresponded generally with indigenous political structure at the time of contact and during prehistory (Milanich 1994). The Apalachee (Boyd et al. 1951; Hann 1988; Swanton 1922, 1946) lived in peninsular Florida between the Aucilla and Ochlockonee Rivers, and their territory circumscribed that of the Tallahassee region and its immediate environs. The Guale lived along the narrow coastal strand of southern Georgia, bounded on the south by the Saltilla River and on the north by the Savannah River (Hann 1987; Jones 1978, 1980; Larsen 1982, 1990; Larson 1969, 1978, 1980; Saunders 2000; Sturtevant 1962; Swanton 1922, 1946; Thomas 1990b). These boundaries are debated (see Hann 1987:2–4; Jones 1978:186; Saunders 2000:15). However,

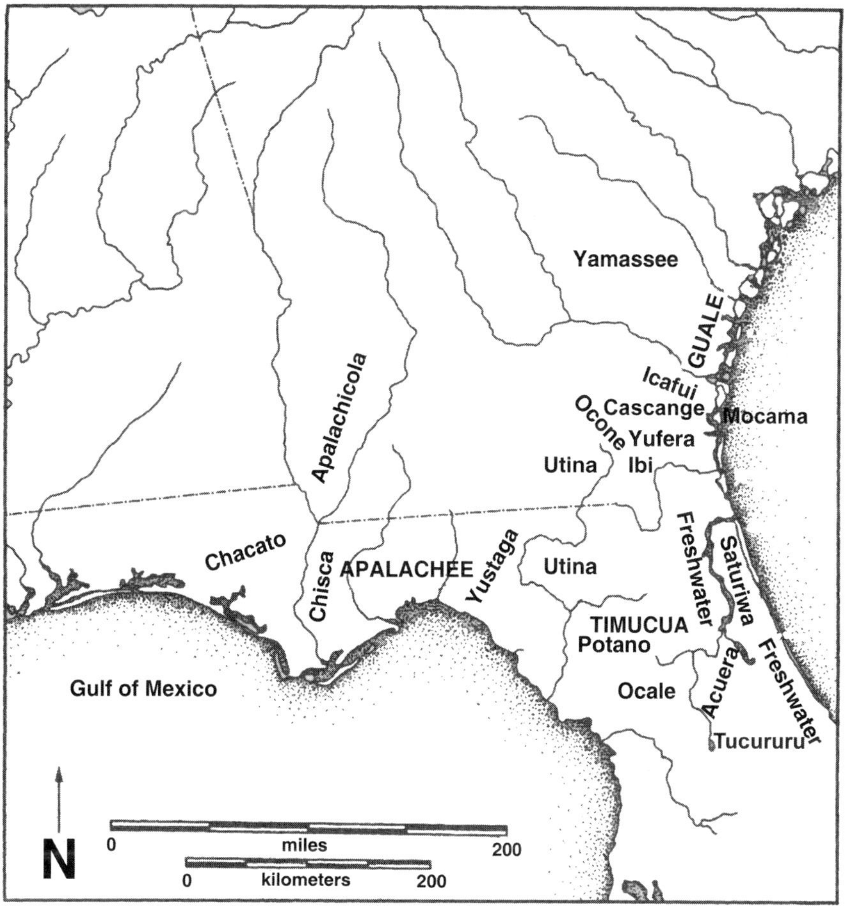

Figure 2.3. Map of mission cultures (capital letters) with individual Timucua-speaking chiefdoms and extralocal, non-Christian communities in lowercase letters (modified from Larsen 2001: Fig. 2.1). Tribe locations based on Hunn (1996: Map 1, p. 2).

the salient fact is that there was clear political differentiation between Guale and groups in the Georgia interior. Timucua speakers occupied a large area of southern Georgia and northern peninsular Florida, directly east of the easternmost Apalachee villages (that is, east of the Aucilla River) to the Atlantic Ocean, with more ephemeral southern and northern distributional boundaries (Deagan 1978; Ehrmann 1940; Hann 1996; Larkford 1984; Milanich 1978, 1996; Milanich and Sturtevant 1972; Swanton 1922; Worth 1992, 1998a, b).

Table 2.2 summarizes significant criteria used to characterize and distinguish these populations. Most critical to this discussion is recognition that the provincial labels (Guale, Apalachee, Timucua) are not equivalent terms. Guale and Apalachee Provinces corresponded with recognized sociopolitical entities

Table 2.2. Comparative summary of mission populations

Feature	Apalachee	Guale	Timucua
PC pop. Size	30,000	~6,000	- 150,000
Pop. size 1700	8,000	~ 200	~ 1000
Pol. organ.	Complex chiefdom	Simple chiefdom	Simple chiefdom(s)
No. major towns	~40	~40–50	~ 200
Pop. density	Dense, permanent villages	Dispersed towns, seasonal mobility	Dispersed towns, seasonal mobility
Structure	Hierarchical	Hierarchical	Hierarchical
Mobility	Sedentary	Semi-sedentary	Semi-sedentary
No. languages	1	1?	11+
Language family	Muskoghean	Muskogean?	Non-Muskogean
Subsistence	Intensive maize agriculture	Horticulture + gathered foods	Horticulture + gathered foods
Soc. org.	Ranked clans	Ranked clans	Ranked clans
Marriage	Polygyny, most monogamous	Polygyny, most monogamous	Polygyny, most monogamous
Marr. restrictions	Clan exogamy	Clan exogamy	Clan exogamy
PMR	Matrilocal	Matrilocal	Matrilocal
Descent	Matrilineal	Matrilineal	Matrilineal
PC cemeteries	Platform mound + homestead	Low mound	Low mound
PCT cemeteries	Mission church	Mission church	Mission church
Max. no. missions	14	12	25–35

Note: PC = Precontact, PCT = Postcontact, PMR = Postmarital residence

(Hann 1988; Milanich 1999). They were both single chiefdoms[1] of roughly equivalent geographic area that were internally culturally homogenous and internally spoke the same language (Hann 1987, 1988; Milanich 1999; Scarry 1992; Thomas 1988). Both chiefdoms can be considered real entities, political units with saliency in regional political affairs. Timucua Province, however, was a large linguistic, not political or cultural, designation that comprised approximately 15–20 distinct independent chiefdoms that were divided into eastern and western components based on size (the eastern groups tended to be smaller) and on subsistence strategies (Deagan 1978; Hann 1996; Milanich 1978, 1996, 1999). The eastern Timucua were composed of ten distinct chiefdoms. The Cascangue, Icafui, Ibi, Yufera, and Oconp lived in southeastern Georgia, while the Saturiwa, Freshwater, Acuera, and Tucururu lived along the banks and tributaries of the St. Johns River (Deagan 1978; Hann 1996; Milanich 1978, 1996, 1999). The Tacatacuru (Mocama) lived on Cumberland Island. The western Timucua were composed of four tribes (Potano, Yustaga, Utina, and Ocale) that lived in the relatively homogenous interior of north-central Florida (Milanich 1978). The Onatheaqua may have been a fifth western Timucua tribe; however, it is uncertain how distinct they were from the Utina (Hann 1996). These tribes spoke approximately 11 different languages that were similar enough to belong to the same Timucuan language family (Hann 1996) but clearly distinct from neighboring Muskogean groups (see Granberry 1993). Therefore, while "Apalachee" and "Guale" are sociopolitical constructs, "Timucua" is a linguistic category that consisted of multiple semi-autonomous chiefdoms, each of which was roughly comparable either to Apalachee or, perhaps more appropriately given Guale's internal political structure, to Guale (see Jones 1978).

Comparison of features listed in Table 2.2 reveals marked similarities between these cultural groups. And, where differences did exist, they were primarily of degree, not of kind. Most similar were aspects of social organization (Hudson 1976). Individuals were grouped into ranked, exogamous, matrilineal clans (Deagan 1978; Hann 1988:70, 1996:82, 112; Jones 1978; Milanich 1978; Oré 1936:107; Swanton 1922). Polygyny was preferable (Barcía 1951:182; Ehrmann 1940:185; Geiger 1937:9, 74, 226; Lanning 1935:30; Laudonnière 1975: 13; Oré 1936:74, 84, 101–103; Zubillaga 1946:418), though likely restricted to elites, and postmarital residence was matrilocal (Hann 1988:70, 1996:113; Larson 1978:126). Gender roles and patterns of labor division were also similar throughout the region.

Differences of degree were recorded for numerous aspects of these populations, however. Subsistence adaptation varied considerably. Although all groups consumed some maize, its importance in the diet of precontact populations generally increased as one moved westward from the coast (Deagan 1990a, b). Eastern Timucuan groups consumed the least maize and subsisted primarily on gathered riverine and marine resources (Deagan 1978; Hann 1996). The Apalachee consumed the most maize and were dedicated intensive agricultural producers of this commodity (Hann 1988). The Guale and western Timucua groups were intermediate with respect to subsistence strategies; however, the coastal Guale obviously made greater use of the resources available from their immediate environment (Jones 1978; Larson 1969, 1978, 1980).

Coincident with differences in maize consumption was similarly patterned variation in the degree of sedentism. The Apalachee were fully sedentary farmers (Hann 1988), while the Guale and eastern Timucua were more seasonally mobile (Deagan 1978; Ehrmann 1940; Hann 1996; Jones 1978; Larson 1978; Laudonnière 1975). The western Timucua may have been intermediate between these extremes (Deagan 1978; Milanich 1978). Although individuals in all three groups were distributed in a hierarchical settlement pattern, settlement density generally increased in concert with the importance of maize agriculture. Apalachee was the most densely and hierarchically settled region (Hann 1988; Thomas 1990b), whereas Guale and Timucua were less so (Deagan 1978; Milanich 1978, 1999). Complexity of political organization was also similar between groups, with evidence for a more complex system among the agricultural Apalachee, who were the only true Mississippian population in Florida at the time of contact (Hann 1988).

The degree of shared similarities should not be construed as an absence of significant cultural variation, however. The populations were linguistically distinct, and although both Apalachee and Guale likely spoke Muskogean languages (Thomas 1990b), they were not mutually intelligible (Hann 1988:120, 1996:125). Different Timucua languages may have been partially mutually intelligible (see comparative lexical homologues in Granberry 1993: Table 4; Hann 1996:7) but were clearly distinct at a macro-taxonomic level from neighboring populations.[2] Linguistic differences did manifest in terms of self-identity and ethnicity, with all three provinces maintaining distinct individual identities well into the 18th century (see Worth 1995:47 for discussion).

In general, protohistoric polities engaged in perennial warfare with neighboring communities. The Apalachee were at war with the Potano, Utina, and Yustaga (Barcía 1951:77; Hann 1996; Steinen and Ritson 1996:13) and with the Chacato (Swanton 1922:119). The Guale were at war with the Mocama and the Orista (Barcía 1951:113; Geiger, 1937:109; Hann 1996:70, 145; Jones 1978:204). In addition to conflict that existed between Timucua speakers and their Muskogean neighbors, warfare was common between Timucuan chiefdoms, which were commonly aligned into fragile confederacies (Deagan 1978; Hann 1996; Laudonnière 1975; Milanich 1978, 1996, 1999). At the time of contact, the major alliances among the Timucua were led by Saturiwa, Outina, Yustaga, and Potano, with other groups figuring less prominently in macroregional political affairs (Hann 1996). The historical record is replete with references to conflict within and between these competing confederacies and chiefdoms (Barcía 1951:77; DePratter 1991:49–56; Elvas 1907; Hann 1996:41–42; Laudonnière 1975:11, 76–77; Milanich 1996; Oré 1936:114–117; Ranjel in Bourne 1904; Steinen and Ritson 1996:112), a fact that may have hindered initial mission efforts in some regions (Hann 1988).

Pattern and Strategy of Mission Expansion

Throughout the New World, Spanish interests were promoted at the expense of the indigenous populations. Because mining and agricultural wealth were the primary economic motives for New World exploration and conquest, Spain readily acknowledged the need for a cheap labor supply. To fulfill this need, they initiated the reorganization of the resident populations in two ways. *Población* was the process by which new areas were forcibly settled by populations from other areas (Bushnell 1990). *Congregación* was the process of congregating dispersed populations into a central location that suited Spanish interests. A special term, *reducción,* referred to congregación populations "without reason" or "without habitation" (Bushnell 1990), the 16th-century Spanish term for semisedentary groups.

Unlike the reducción system that was practiced in South America and the Caribbean, the Franciscans did not *initially* force the aggregation of dispersed populations into new, congregated settlements (Deagan 1978, 1985, 1990a, b; Hann 1986b, 1988; Sturtevant 1962; Thomas 1990a). Rather, through identification of local political landscapes, the missionaries managed to establish primary mission locations at the village of the local paramount chief. By their working in concert, successful conversion was facilitated (Thomas 1990a).

The primary mission village, or *doctrina,* served as the residence and central meeting place for converts. The date of the doctrina's founding was typically commemorated by the first half of its name. The second half recognized the principal chief or village in which the doctrina was located. For example, San Lorenzo de Ivitachuco was established on the day of Saint Lorenzo in the village named or ruled by Ivitachuco.[3] In addition, friars made rounds to outlying villages (*visitas*) to administer daily rites. One obvious effect of this system was that satellite villages under a mission's jurisdiction needed to be within a reasonable walking distance. For Apalachee Province, Governor Rebolledo's visitation record indicates that primary villages had two to four subordinate satellite villages (Hann 1986a; Marrinan 1993), whereas San Martín de Timucua had between 5 and 20 villages under its jurisdiction (Weisman 1992:34). The microlevel structure of an individual doctrina, therefore, varied considerably and likely reflected both the political importance of the primary chief and the size and density of the local population.

Initially, the sphere of influence of the mission centers generally corresponded with pre-existing political and social networks (Deagan 1978, 1985,

1990a, b; Hann 1986b:371, 1988:102; Worth 1992:40), a fact supported by archaeological evidence for precontact components at some mission sites (Deagan 1985; Marrinan 1993).[4] The forced population aggregation experienced in other regions of the New World was the direct result of the need for a reducción and encomienda system to expedite the extraction of mineral or agricultural resources. *La Florida* was economically poor, and encomienda had been outlawed by the time of St. Augustine's founding (Deagan 1985; Hussey 1932). It was only after severe demographic collapse seemed imminent that directed aggregation and reorganization of the Florida and Georgia populations was adopted as a Spanish strategy for the maintenance of their colony (Worth 1995).

Franciscan reliance on pre-existing political and economic institutions resulted from their limited military and subsistence support (Ehrmann 1940). In addition to political alliances, the friars also increased success by embracing elements of indigenous culture while simultaneously preaching the Christian faith. One means of achieving this combination was instructing the Indians in their native language. In addition, in an effort to avoid conflict, the friars accepted customs not deemed detrimental to a Catholic lifestyle; however, attempts were made to give such customs a "Christian interpretation" (Geiger 1937:30). Practices contrary to Christianity, for example, polygyny and the playing of the ballgame with its overt pagan symbolism, were targeted for eradication. Franciscan philosophy was aptly summarized by Geiger (1937:30): "while every attempt was made to make a good Spaniard out of the Indian, the missionaries were more than satisfied if they made a good Indian out of him."

Bajo Campaña: Life Under the Bell

Despite delineation of a "Golden Age" of the Florida mission period (Geiger 1937), others have taken issue with the overall success of the venture (Spellman 1965) in recognition that this misnomer belies the harsh reality of the mission experience. Although the missions existed for approximately 150 years, problematic relations with indigenous authority, conflict between religious and secular officials, and economic hardship were common. The Spanish Crown is often accused of ignoring the colony, resulting in late or insufficient subsidies (Bushnell 1978a; Spellman 1965; but see Sluiter 1985).[5] Tension between the friars and the military was also problematic, and political corruption drained the already weak economy. At its core, the economic and political structure of the *La Florida* colony was troubled (see Bushnell 1994).

Three-way tension among the friars, the secular Spanish government, and the indigenous elite fomented frequently into periods of active resistance and revolt: in 1576, 1587, 1597, 1608, 1645, and 1680 for Guale (Saunders 1992, 2000; Swanton 1946); in 1656 for Timucua (Worth 1992); and in 1647 for

Apalachee (Hann 1988). One tried-and-true method of ameliorating unfavorable living conditions or the unreasonable labor demands of *repartimiento* (a system of labor tribute) was fugitivism (Bushnell 1990; Spellman 1965; Worth 1998a). The disenfranchised simply left their natal villages to become "ghosts" in the system, a practice that exacerbated population losses that resulted from English-sponsored slave raiding and, above all, from epidemics (Worth 1998b).

The impact of European-introduced diseases on the indigenous populations was staggering in some areas. A partial list of major recorded epidemics includes the years 1526, 1570, 1582, 1586, 1591, 1613–17, 1649, 1653, 1657, 1659, 1670, 1672, 1675, 1686, and 1703 (Bushnell 1982:13; Deagan 1978:94, 1985; Dobyns 1983, 1991; Hann 1988; Larsen 2001; Swanton 1922:337). While increased mortality rates are an obvious sequelae, paleodemographic models suggest that demographic collapse was hastened by concordant declines in fertility. Russell (1987) examined age-at-death profiles for the Guale bioarchaeological record and found that postcontact populations experienced a lower mean age-at-death and decreased survivorship in comparison to late precontact populations, a conclusion tempered by concern over sampling bias and infant underenumeration (Russell et al. 1990). More recent research by Larsen and coworkers (2001) utilized two parameters: mean age-at-death (Sattenspiel and Harpending 1983) and the 30+/5+ age ratio (Buikstra et al. 1986) to investigate changes in fertility commensurate with documented increases in mortality. Although pooling of samples is problematic, the results generally indicate a decline in fertility with incipient missionization, followed by a slight increase in fertility during the later phases of the mission period. Results of the paleodemographic analyses did suggest that individual missions experienced different fertility rates, however.

Three features of mission life helped catalyze deteriorating living conditions. First, the Spanish successfully increased emphasis on maize production, a product they viewed as a valuable commodity. Increasing consumption of maize by indigenous populations further worsened deteriorating health conditions. Second, settled village life in aggregate communities fostered additional vectors for disease transmission and additionally selected for environments characterized by poor sanitation and increased morbidity. Third, implicit with conversion was loyalty to the Spanish Crown. The implied submission instituted significant physical demands on the native populations, for example, the planting of communal fields, carrying loads between missions, and contributing to communal labor projects in St. Augustine. Bioarchaeological analyses have successfully documented how harsh life "under the bell" may have been for the converted (Larsen 1993, 2001). Because many of the deleterious effects of aggregate mission life can be related to dietary changes, I begin the discussion here.

Dietary shifts have been detailed in numerous bioarchaeological analyses using both macroscopic observations and bone chemistry signatures. It is well known that a high carbohydrate diet (such as one with maize) leads to decreasing oral health and increased risk for caries development. Caries (or cavities) are areas of enamel demineralization that result from the acidic by-products of specific oral bacteria that metabolize ingested carbohydrates (Hillson 1996). For the Florida record the evidence is unequivocal. Caries prevalence is lowest in the precontact, pre-agricultural populations, increases differentially during the late precontact, agricultural period in concert with differential commitment to maize agriculture, and then uniformly increases for all regions after missionization (Larsen et al. 1991, 2001). Variability during the late mission period adds a degree of complexity to the interpretation (Larsen and Tung 2002); however, the general trend is nonetheless well established.

Stable isotope analyses contribute to discussions of dietary shifts by focusing on two elements: Nitrogen (δ^{15N}) signatures decrease as marine food resources become less important and carbon (δ^{13C}) signatures increase (become less negative) as maize consumption increases. Regional perspectives have outlined several broad patterns (Hutchinson et al. 1998, 2000; Larsen et al. 1992, 2001; Schoeninger et al. 1990). Precontact, pre-agricultural populations demonstrate limited evidence for maize consumption with marine contributions varying by local diet, as expected. After A.D. 1000, maize agriculture appears throughout the region but is differentially incorporated into local diets. For the Apalachee the transition is marked, for the Guale the transition is noticeable but less severe, and for the coastal Timucua the effect of maize is negligible. After contact, however, dietary strategies homogenize. Those populations that consumed maize during precontact times (Apalachee, Guale) became more committed to this staple, and those populations that did not consume much maize began to do so (Timucua). Nitrogen signatures suggest that where marine foods were locally available during the precontact period, they were consumed. However, the uniform decrease in nitrogen isotopic signatures after contact, combined with the increase in values of carbon isotopic signatures, suggests that maize became more important while marine resources became less important (Larsen 2001).

Homogenization of diets throughout the Florida colony is reflected in patterns of incisor and molar microwear (Teaford et al. 2001). As with isotopic signatures, the patterns of pits and scratch orientations on the dentition suggest significant local variation, which is partially related to differences in soil composition (clay vs. sand). Greater homogeneity of microwear features during the precontact period suggests homogeneity in food preparation strategies. More variable scratch orientations for postcontact populations may indicate a decline in toughness of the diet; consumption of processed maize products (gruels) requires less precision processing, resulting in more chaotic and varied microwear orientations. In this case, cross-regional consistency in terms of greater scratch orientation variability indicates the adoption of similar maize-based diets throughout the *La Florida* colony.

Ezzo et al. (1995), however, disagree that maize replaced marine foods, at least for the Guale. Using strontium (Sr) and barium (Ba) trace elements analyses, Ezzo and colleagues documented low levels of barium and moderate levels of strontium in the contact period populations. In addition, the Ba/Sr ratio decreased throughout the mission period, suggesting an increase rather than a decrease in marine resource utilization. The authors reconcile their results with those of bone isotope analyses by proposing that maize did not replace marine foods in the diet but rather replaced other wild plant foods. Therefore, marine resources remained important in the diet, but plant consumption changed from a diverse repertoire of wild plants to a heavy reliance on maize alone. The overall conclusion remains the same, however. Dietary breadth decreased during the contact period (Larsen et al. 1996, 2001), and population health suffered as a result.

Caries incidence is but one indicator of declining health conditions. And, while in the modern era caries may seem more of an expensive inconvenience, oral health is quickly becoming recognized as a significant predictor of the overall health of an individual. With poor oral health comes additional health concerns. Analysis of the effect of missionization on contact period health has additionally utilized information on linear enamel hypoplasia (LEH) prevalence, frequency, and duration/severity (Figure 2.4), pathological striae of Retzius, cribra orbitalia (Figure 2.5), and porotic hyperostosis and general periosteal bone infections (Figure 2.6).

In combination, these macroscopic pathological morbidity indicators suggest a decline in health during the mission period related to a decrease in dietary diversity, increased metabolic stress, and increased parasitic infections. One of the more commonly recorded markers of physiological stress is linear enamel hypoplasia of the permanent dentition. These enamel defects represent periods of growth disruption during amelogenesis (Hillson 1996). Mission bioarchaeologists have now examined the frequency and severity of these growth arrest markers for approximately 2,000 teeth representing 772 individuals from 34 skeletal samples (Hutchinson 1986; Hutchinson and Larsen 1988, 1990, 2001; Larsen and Hutchinson 1992; Storey 1986). The results present a complex picture of biocultural adaptation but generally indicate that the late mission period was more stressful than the early mission period and that health declined steadily throughout the period of missionary activity (Hutchinson and Larsen 2001). These conclusions are tempered by the fact that contact period samples generally exhibit fewer stress events than their precontact ancestors, indicating a complex etiology to population expression.

Simpson (Simpson 2001; Simpson et al. 1990) recorded information on the prevalence of pathological striae of Retzius in 143 individuals from 14 sites

Figure 2.4. Example of hypoplastic defects (white arrows) on a lower canine, giving an extremely uneven appearance to the labial surface of this tooth. Note also the excessive calculus deposit near the cemento-enamel junction that precludes accurate assessment of the buccolingual diameter of this tooth. Photo courtesy of Colette Berbesque.

dating to the precontact and mission periods. These defects in crown formation represent short periods of stress, likely related to infant diarrhea events resulting from parasitic infection whose modal distribution of occurrence is between 12 and 30 months of age (Simpson 2001). Because these markers occur much earlier in life than surface hypoplastic defects, and represent much shorter episodic perturbations, Simpson proposed a distinct etiology and period of onset (roughly the first through third years of life as opposed to hypoplastic defects that cluster around the third through fifth years of life).

Simpson reported two statistics of interest: percentage of individuals per sample that exhibited at least one pathological striae of Retzius and, for afflicted individuals, the average number of pathological striae. His results can be summarized as follows: for Florida early precontact, 67 percent afflicted with an average of 1.8 insults per individual; for Florida late precontact, 36 percent afflicted with an average of 2.6 insults per individual; for the Ossuary at Santa Catalina de Guale de Amelia, 54 percent afflicted with an average of 1.7 insults per individual; for early mission samples, 83 percent afflicted with

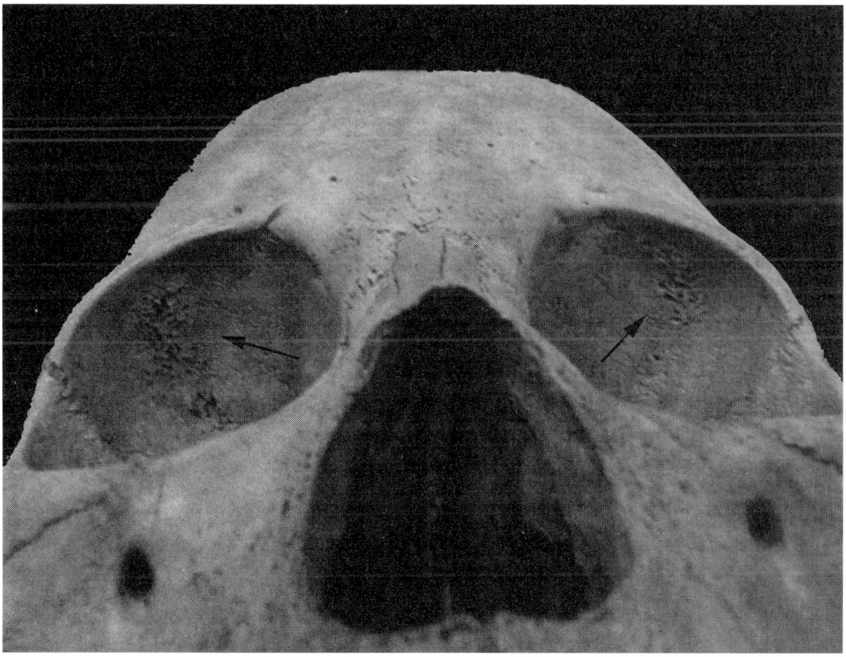

Figure 2.5. Example of cribra orbitalia in a prehistoric Florida native. Note the lesions evident in the horizontal portion of the frontal bone (arrows) forming the roof of the orbit. These lesions result from the expansion of the diploë through the outer table of cortical bone that forms the roof of the orbit.

an average of 2.1 insults per individual; and for late mission samples, 82 percent afflicted with an average of 2.4 insults per individual. Overall the prevalence of pathological striae increased from 48 percent affected before contact to 83 percent afflicted after contact, despite limited changes in the number of insults per individual (2.2 vs. 2.5 insults, respectively). Simpson's research indicates that the effects of marginal health conditions were particularly evident for the earliest years of life.

Skeletal indicators of pathological conditions likewise indicate that missionization had negative health consequences for mission communities. Cribra orbitalia and porotic hyperostosis both increase significantly in prevalence during the mission period (Schultz et al. 2001). This suggests that iron-deficiency anemia was common in mission communities and may have resulted from dietary deficiencies, overcrowding and subsequent parasitic infection owing to poor sanitation, or a combination of both (Larsen and Sering 2000). Periosteal bone infections of the tibia and femur similarly increase in frequency after contact (Larsen and Harn 1994).

In addition to changes in population health conditions, bioarchaeological

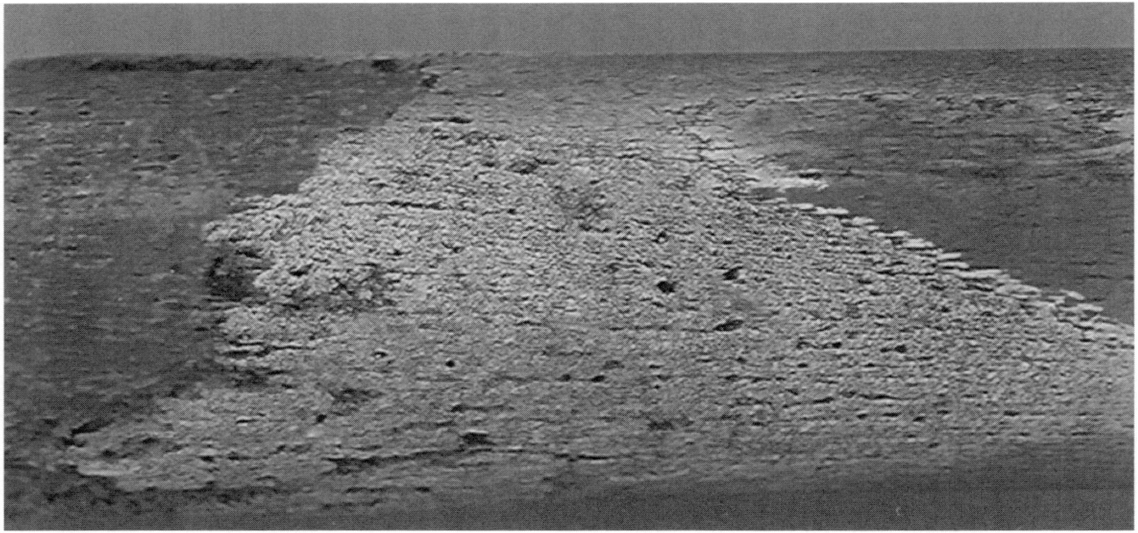

Figure 2.6. Example of a periosteal lesion from a prehistoric Florida native.

data indicate that behavioral activity patterns changed and physical stress increased after contact. The pattern and prevalence of osteoarthritic lesions indicate an increase in frequency of affected bones during the contact period. The most dramatic changes occurred in the vertebral joint surfaces, with males exhibiting a higher frequency of arthritic modifications (Griffin and Larsen 1989; Larsen et al. 1996), suggesting an increase in the mechanical demands placed on the back, typical of lifting heavy objects. That males were enlisted as cargo bearers in the repartimiento system may explain the sexual dimorphism observed in pathological joint modification patterns (Griffin and Larsen 1989).

Fresia and colleagues investigated sexual dimorphism in the mechanical use of the upper arm for the Georgia coast populations (Fresia and Ruff 1987; Fresia et al. 1990). Using radiographs of the humeri and calculations of bone area and second moments of area, they demonstrated that bilateral asymmetry of the humerus decreased through time. This finding indicated that males and females were performing more similar kinds of upper arm activities during the contact period. Because females displayed the largest change from the hunting and gathering to agricultural periods, and because males exhibited the largest change with the beginning of the mission period, they concluded that the decline in sexual dimorphism resulted from males having assumed a more prominent postcontact role in maize production.

Ruff and Larsen (Larsen and Ruff 1994; Ruff and Larsen 1990, 2001) investigated changing activity patterns among the Guale using several biomechanical models of cross-sectional longbone geometry. Relative strength measures indicated an increase in femoral strength, particularly in females, an increase in humeral strength for males, a decrease in humeral strength for females, and an overall increase in body size during the contact period. The shape ratio of the mid-femur (I_x/I_y) declined in females but increased in males, suggesting that long-distance travel demands increased for *some* males. Both sexes demonstrated a trend toward subtrochanteric noncircularity in limb dimensions, suggesting a decline in overall activity levels (Larsen and Ruff 1994; Ruff and Larsen 1990). In summary, the biomechanical evidence combined with pathological joint alterations indicated that the mission period was characterized by heavier work loads, increased body size and weight, and, for some males, increased long-distance travel.

The *La Florida* mission system provides an ideal setting in which to test models of Native American responses to impending demographic catastrophe and European hegemony. Surviving for almost two centuries, and beginning years before the more visible Spanish missions of the American West, the Florida colony united myriad cultural and linguistic groups under the banner of the Cross. Apalachee, Guale, and multiple Timucua-speaking populations received the friars, sometimes openly, sometimes with reservation. After peak mission expansion during the first half of the 17th century, life for the mission Indians was difficult. Economic hardship, political strife, and disease epidemics instigated periods of unrest. Native revolts and fugitivism hastened demographic collapse in *some* communities. Forces external to the mission system and the impending European power struggle over control of North America combined to paint a dismal picture of mission life that included increased physical stress and morbidity, decreased dietary breadth for most populations, and increased parasitic infection.

This chapter was concerned with placing this book within previous regional scholarship. Having generated an outline of significant events, personalities, and lexicons, I am now in the position to provide specific ethnohistoric details to generate an evolutionary research design. Genetic drift and gene flow (in the biological sense) correspond with population size and population migration (in the ethnohistoric sense). Chapter 3 discusses general processes of demographic collapse and population aggregation with the goal of providing comparative (Apalachee, Guale) research proposals.

Notes

1. Jones (1978) has identified three subchiefdoms affiliated with the Guale: Guale-Tolomato, Asao-Talaxe, and Espogache-Tupiqui.

2. The earliest to consider the linguistic affiliation of the Timucua language family was Swanton, who initially viewed it as a language isolate (Swanton 1922) but then came to appreciate similarities between Timucua and the Muskogean languages (Swanton 1929). Although Swanton himself described these similarities as structural rather than lexical, Granberry considers all Muskhogean elements to be loan words from a proto-Muskogean stock (Granberry 1993). Swadesh (1964) noted similarities to Arawakan languages, and Haas (1971) categorized the affiliation of the language as unknown though possibly Souian in origin. The most complete analysis of the language was conducted by Julian Granberry, who inherited Swanton's original documents on the subject. According to Granberry, "If we have difficulty in finding a dominant lexical contributor to Timucua, we have, as suggested above, less trouble in finding a primary grammatical contributor. The basic patterns of Timucua grammar conform rather closely to Macro-Chibchan. The nearest similarities are, on the other hand, to the Warao isolate of the Orinoco Delta in far eastern Venezuela. There are also individual morphemes and lexemes with striking resemblance to modern Warao as well as an even larger number of lexemes with equally striking resemblance to languages of the Vaupés-Caquetá-Inirida-Guaviare branch of Northern Maipuran Arawakan. The number and close correspondence of nominal and verbal prefix suffix morphemes with those of Warao is noteworthy—44% of Warao noun suffixes and 17% of Warao verb suffixes have Timucua parallels or identities" (Granberry, 1993:15–16).

3. Garcilaso reported that chiefdoms, primary towns within the chiefdom, and the chief himself frequently were conferred the same name (Varner and Varner 1951:122).

4. According to Deagan the following missions had precontact components: San Juan del Puerto, Nombre de Dios, all Apalachee missions, Santa Catalina de Guale, and the Tacatacuru mission (Deagan 1985:303). However, both Baptizing Spring and San Martín de Timucua appear to represent new, previously unoccupied locales.

5. Sluiter (1985) examined 80 years of original Spanish records and found no evidence of a single late or canceled payment.

3 Bioethnohistory

> When a king dies, they bury him very solemnly, and on his grave they put the cup from which he usually drank. All about the grave they stick many arrows, and there are three days and three nights of continuous fasting and lamentation. All the kings who are friends of the departed king mourn in the same manner. As testimony of their friendship to the deceased, they cut off more than half their hair, men and women alike. For a period of six months certain women are delegated to mourn the death of the king, crying out in a loud voice three times each day, morning, noon, and night. All the worldly goods of the king are put in his house, which is then set on fire so that nothing of his will ever be seen again. They do the same thing for their priests; and a priest's body is put in the house before it is set afire.
>
> (Laudonnière 1975:14–15)

Exploration of regional ethnohistories is a fascinating endeavor for a biological anthropologist. Paleographers, those doing the direct translation of original source material, do not have vested interests in evolutionary research questions, nor might they know what to look for if they did. Because I myself am not a historian, let alone a paleographer, I cannot add substantially to the critical discussion of the translations. However, I can contribute to the discipline by outlining the kinds of information that paleogeneticists find useful for generating research designs and addressing interesting questions about the past.

This chapter distills the voluminous historical material into several basic categories that human biologists would find most useful for hypothesis formation. Using these historical data, I formulate specific evolutionary theoretical expectations in chapter 4.

Previous Research

In a series of papers (Griffin 1989, 1993; Griffin and Nelson 1996; Griffin et al. 2001), Griffin and colleagues analyzed phenotypic data from 13 archaeological skeletal samples dated to the period A.D. 1200–1700. Three of the sites date to the mission period (Santa Catalina de Guale, Santa Catalina de Guale de Santa María, and Santa Maria de los Yamassee) and are affiliated with the coastal Guale, Yamassee, or Timucua. Comparative data include the north-central Georgia coastal Irene Mound series, two coastal North Carolina samples, and seven samples from the interior of Georgia, North Carolina, and Tennessee.

Griffin's data set consisted of 35 dental morphological variables and 25 cranial nonmetric features (see Griffin et al. 2001: Tables 9.2, 9.3) that were subjected to a series of multivariate statistical analyses. Unfortunately, the cranial and dental data (which were analyzed separately) did not always produce consistent results. Given the generally lower reliability of cranial nonmetric variables (compare Sjøvold 1984 with Scott and Turner 1997), I place greater emphasis on the results based on dental morphology. The conclusions of Griffin and colleagues can be summarized as follows.

(1) Despite ethnohistorical and historical indications to the contrary, there was a clear biological distinction between the Guale mission samples and the Irene Mound series. This distinction is surprising, given the close geographic proximity of these sites, and suggests that a biological discontinuity existed between the late precontact and early historic period populations in coastal Georgia (Figure 3.1).

(2) The historic period samples demonstrated close biological affinity, which is unexpected, given the supposed distinction between the Guale and Yamassee ethnic groups (Bushnell 1986). The distinction between Santa Catalina de Guale and Santa Catalina de Guale de Amelia Island was greater than might be expected, given the documented historical continuity between these samples (Larsen 1993; Saunders 1993), but it was consistent with expectations based on differences in population size and the effects of random genetic drift.

(3) The pattern of phenotypic variability suggests that the early mission period was characterized by either extensive gene flow between diverse groups (including inland populations) or population aggregation at the early mission period Santa Catalina de Guale site. The late mission period (post-1686) Santa Catalina de Guale de Amelia sample demonstrated considerably less phenotypic variability in comparison to Santa Catalina de Guale. This difference is illustrative of the demographic collapse that characterized coastal populations in the later 17th century and represents the effects of the genetic bottleneck that must have differentiated the Santa Catalina and Amelia Island populations. This result forms the basis for the research questions evaluated in this book.

Bioethnohistory

I use ethnohistoric data to model two evolutionary processes, namely gene flow and genetic drift (as discussed in chapter 4, mutation and natural selection figure little into my modeling). Ethnohistoric data provide two contributions to studies of gene flow in extinct populations: establishment of population boundaries that existed at some time in the past, and information on how these boundaries may have changed through time. Ethnohistoric data also contribute to discussions of population size, a parameter which figures directly into modeling the effects of genetic drift on levels of population genetic variability. As population size decreases through time, the expected effects of genetic drift

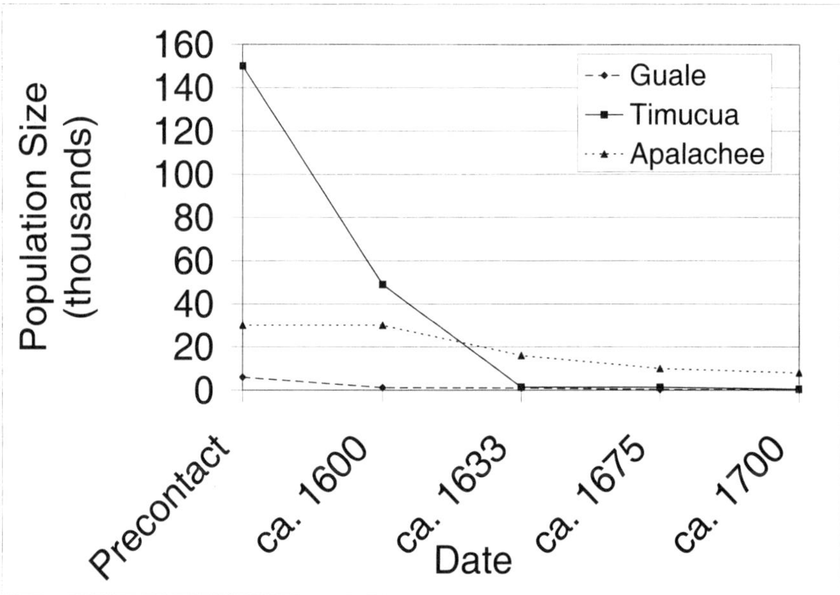

Figure 3.1. This figure presents general consensus population size estimates for Guale (diamond, dashed line), Timucua (square, solid line), and Apalachee (triangle, dotted line) for five time periods. Precontact estimates are most contentious, whereas postcontact estimates are based on enumeration data from Hann (1988, 1996), Worth (1995, 1998b), Milanich (1978, 1996, 1999), Lanning (1935), and Deagan (1978). The conflation of Timucua as a category similar to Guale and Apalachee obscures reasonable estimation of effective population sizes for individual Timucua polities, which results in a swamping effect in this figure. The decline in Guale population size looks less precipitous due to the scale required to accommodate the large Timucua precontact estimate.

increase (Relethford 1996). Both the probability of allele fixation (leading to a reduction in heterozygosity) and the rate of allele fixation (how many generations this takes to occur) are affected by the effective population size (Hartl and Clark 1987). Parameter estimation for gene flow and genetic drift cannot be parceled into distinct components; one cannot consider changes in population size without first defining the limits of the population. Once defined, however, ethnohistoric data can be used to estimate effective population sizes for specific time intervals and then to reconstruct how population size changed through time.

The challenge in combining evolutionary and ethnohistoric data lies in incorporating general principles with very specific ethnohistoric facts. Ethnohistoric data provide statements that biologists can use to model evolutionary processes in the following way.

1. Define the populations and the population structure (systems of mate ex-

change) during the late precontact period. Seven classes of ethnohistoric data contribute to this discussion: (a) political structure; (b) warfare and conflict; (c) linguistic diversity; (d) mobility and sedentism; (e) trade and exchange patterns; (f) marriage patterns and social organization; and (g) direct evidence for interethnic intermarriage. These data can be combined to generate a model of the degree of intermarriage or interbreeding between villages within a province, between provinces, and between provinces and groups external to the mission system.

2. How population size changed through time: Several events/processes lead to reductions in population size within mission communities (Worth 1998b): (a) slave raiding; (b) fugitivism; (c) death due to marginal mission health conditions and Spanish-induced hardship (repartimiento labor); (d) epidemic disease; and (e) reduced birth rates. Factors that affect population size also initiate changes in population boundaries leading to the next point.

3. How population boundaries changed through time: This refers to contact period processes, largely a reaction to changes in population size. Ethnohistoric evidence of interest includes (a) aggregation of local villages to the doctrina (Stage 1 congregación); (b) aggregation and consolidation of doctrinas (Stage 2 congregación); (c) wholesale replacement and relocation of populations (reducción); (d) the demographic "vacuum" and in-migration from extraregional sources (reducción); (e) societal breakdown leading to restructuring of mate exchange proscriptions. Both new mating opportunities and necessity may play a role in this restructuring. Depending on the particular structure of populations at the time of contact (as delineated from 1a–g above), changes in mate exchange patterns could result from (1) consolidation of villages within a province; (2) migration between provinces; (3) in-migration from populations external to the mission system (for example, Creeks, Chisca, Yamassee); and (4) admixture with European colonists or with African slaves.

Whereas the expected phenotypic effects of consolidation (3.e.1), migration (3.e.2), and in-migration (3.e.3) are dependent on precontact population definition and structure (1.a–g), admixture with European/African populations (3.e.4) would certainly increase within-population variability because it is known that these populations were distinct before contact. Therefore, any evidence of European or African admixture would predict an increase in within-population genetic variability. Processes 3.e.1, 3.e.2, and 3.e.3 would lead to an increase in phenotypic variability *only* if the populations were previously distinct and partially biologically differentiated before contact. If this was the case, then genetic drift would have, over time, led to increased genetic variances between populations and reduced genetic variances within populations. If the microregional structure is subsequently altered, between-group differences in allele frequencies would disappear, resulting in reduced between-population genetic variances but, over the short term, increased within-population genetic variances. The preceding is predicated on the assumption of initial (time = 0) conditions of genetic isolation. Only then would admixture between groups increase levels of within-population genetic/phenotypic variation.

The Ethnohistoric Data

Although I have presented the above discussion in a logical progression beginning with the precontact period and continuing through the processes operating during the mission period, it makes greater sense to discuss specific examples in reverse order—that is, to present documentation of changes known to have occurred during the contact period and then discuss hypothesized precontact population structure. The former is less contentious and more matter of fact; the latter, as will be seen, is far more problematic and conjectural. In addition, a majority of the mission-era processes apply across provincial populations (Apalachee, Guale, and Timucua) such that general statements can be offered for the entire study region and then modified accordingly with respect to variation in localized mission experience. Outlining general principles is much more difficult for precontact period population boundaries, however, because of the specificity of the relationships being modeled. For example, it is easy to generalize that population sizes declined for all missionized populations during the mission period. However, predicting the evolutionary effects of Apalachee and Yustaga intermarriage requires a more carefully specified argument.

I therefore structure the remainder of this chapter following the outline generated above, with the exception of moving sections 1a–g to the end of the discussion. The most salient feature of the mission period is demographic collapse, which has direct application for models of genetic drift but also simultaneously provides the impetus for the restructuring processes that affect postcontact boundary definitions. After presentation of these data, the hypothetical effects of demographic collapse on the mission populations are explored.

Evidence for Population Size Decline

Received wisdom dictates that population size declined significantly after European colonization of Florida; however, the mechanism and rate of collapse are debated (Dobyns 1983; Henige 1998; Ramenosky 1987). My purpose in this chapter is not to discuss the issues surrounding New World pandemics but rather to provide some general statements about the rate of population size decline in the mission provinces. As will be seen, my contribution to this discussion suggests that some of these population size estimates may be problematic.

With the exception of a handful of Apalachee descendants still alive today,[1] Florida's precontact indigenous populations have completely disappeared from history. Final numbers by province are debatable, but all ethnic groups were represented by several dozen individuals or fewer after final retreat of Spanish loyalists to the St. Augustine area during the first decade of the 18th century. For example, there were 61 remaining Guale living in a single village near St. Augustine in 1711 (Worth 1995, 1998b:148), with similarly small populations of Apalachee and Timucua speakers (Swanton 1946).[2] Despite documentation of various Apalachee enclaves throughout the interior southeastern United States during the 18th century (Covington 1964; Hann 1988; Sattler 1992), most would consider Moore's 1704 raid to be the death knell for the Apalachee culture; Timucua and Guale disappear outright.

Accepting the near complete destruction of these populations in the wake of disease epidemics and other mission-related hardships (Worth 1998b), the most difficult task is trying to estimate the "virgin soil" population sizes for each province, a debate to which much ink has been devoted (Barcía 1951:163; Boyd 1948; Deagan 1973, 1978:95, 1985; Dobyns 1983:13, 1991; Elvas 1907: 173; Hann 1986a:378–379, 1988:117, 1996:174; Jones 1978:183, 185, 198; Lanning 1935:12, 18, 22, 51; McEwan 1991a; Milanich 1999:45; Sturtevant 1962: 68; Swanton 1922:337, 1946; Zubillaga 1946:418), and then tracking the rate of reduction throughout the mission period. Historian John Worth has aptly summarized the problem.

> Population estimates (on both a local and regional scale) by sixteenth-century European observers are virtually nonexistent, and would be highly suspect even if they existed due to the lack of direct and thorough European reconnaissance in most areas. More importantly, the impact of indirect European contact during the early and mid-16th century, including not only undocumented epidemics but also Euro-Indian conflict (sparked between aboriginal chiefdoms as a result), is virtually impossible to gauge, but was undoubtedly substantial. By the time the first even vaguely reliable population estimates began to appear around the turn of the seventeenth century, pre-European population levels had undoubtedly changed. (Worth 1998b:2–3)

For my purposes, raw numbers are of limited concern. Estimates of proportional reduction in population size are, however, integral for accurate modeling of potential drift effects. The quality of the data is highly variable, as Worth notes, which is compounded by the rigorous requirements of population size estimates for microevolutionary research. All estimates are not equally as useful for evolutionary predictive modeling (Relethford 2003). Historians are generally concerned with total census size estimates that would include adult males and females and children of both sexes. Evolutionary biologists, however, differentiate census population size from breeding or effective population size, the latter figuring directly into models of genetic drift and population structure (Hartl and Clark 1987; Relethford 2003). Although definitions of effective population size can be exceedingly complex (Hartl and Clark 1987; Relethford 2003), a simple example should illustrate the concept.

A typical population size culled from the historic literature might be "1,000 adult males" for a specific mission. Assuming equal sex ratios, 2,000 total adults, roughly 18 to 60 years of age, is a reasonable estimate. To this we add the number of children of nonreproductive age, which, assuming an equal pairing of males to females and two children per marriage, results in the addition of another 2,000 individuals, or a census population size of 4,000. This crude estimate has a number of demographic assumptions built into it: equal sex ratio, maximum age at death, earliest age of reproduction, average number of offspring per couple, and 0 population growth (each couple is simply replaced by two children). Any or all of these are unlikely to be true for the mission period. From an evolutionary perspective, however, even with these demographic assumptions, the estimate of 4,000 individuals grossly overestimates the actual effective size of this population: prereproductive children and postmenopausal females are of limited evolutionary significance because they cannot contribute to patterns of genetic variation in the succeeding generation. This estimate also assumes that the society is monogamous and not inbred (Relethford 2003). The ability of a population to reproduce itself, therefore, varies as a function of the number of reproductive age individuals in the population at one point in time. In my example, if only 10 of the 1,000 females were of reproductive age, then the effective population size would be closer to 40 individuals than to 4,000 individuals. Accepting several assumptions about the underlying age and sex structure of the population, a rough estimate of the effective size is ⅓ estimated census size (for each sex ⅓ are prereproductive, ⅓ are reproductive, and ⅓ are postreproductive), although this proportion can vary considerably within the known range of human adaptive strategies (Jorde 1980; Relethford 2003; Wobst 1974).

Another key point to consider is that raw numbers are of little concern. Proportionate reduction in population size through time (as a percentage) and relative (between-province) population size estimates are of primary interest. And, while demographic transitions during the mission period suggest that the ratio of census to effective population size is changing through time,[3] some of these concerns are ameliorated by assuming that reductions in Apalachee, Guale, and Timucua will be roughly comparable. In other words, the demographic transition roughly affects the age and sex structure within each province in a similar manner.

Vagaries of the historical documents provide additional difficulties. Many of the earliest enumerations that postdate contact but predate the missions consist of counts of villages converted into census sizes based on rough density estimates (Jones 1978; Lanning 1935). Other data are invariably related to warfare: estimates of the size of specific tribal confederacies, for example, Nicholas Bourginon's estimate of 10,000 individuals controlled by the Saturiwa alliance (Hackluyt 1903:113) or estimates of the number of warriors to be marshaled to battle, for example, Groutald's mid-16th-century claim that Yustaga could field 3,000–4,000 warriors (Milanich 1978:64). The nonspecificity of the categories enumerated leads to problems in comparison. The former refers to an abstract confederacy of tribes that quickly disappeared after contact (Deagan 1978) and has a particularly vague referent (is it male warriors, total adults, or total individuals including elderly and the young?), let alone inherent biases in the original source (a Frenchman remarking on the size of an indigenous alliance that the French were actively manipulating). In the latter estimate for the western Timucuan Yustaga chiefdom, there is the difficulty of defining the age class referenced. Assuming "warriors" refers to prime breeding age males (roughly 18–45 years of age), this suggests a census population size of approximately 18,000 to 32,000 individuals.

Sources for mission period estimates become more diverse and derive primarily from either secular or Franciscan enumerations, both of which were subject to bias. The ethnohistoric record has provided estimates of number of males, number of warriors, number of families, number of baptisms, number of confirmations, number of newly confirmed, and number of Christians (see Deagan 1978; Hann 1988: Table 7.1, 1996; Milanich 1978, 1996, 1999; Oré 1936; Wenhold 1936; Worth 1995, 1998a, b). Normalizing these estimates requires assumptions about family size and the age classes included in baptismal and confirmation records (which are subject to considerable exaggeration by the Franciscans). In addition, many of the postmissionization estimates (and by definition all baptismal or confirmation numbers) refer to Christian converts only, which, given the documented intensity of intergroup migration and in-migration, could actually underestimate effective population size considerably. This problem is particularly acute in the late 17th century when non-Christian Indians could comprise 50 percent or more of local population bases with no evidence for or against potential breeding effects recorded (see Worth 1995).

Establishing precontact population size estimates for Guale is particularly fraught with difficulty, owing in part to poor definition of the geographic extent of the territory. While the southern boundary seems to coincide well with the distribution of Irene ceramic types (north of the Altamaha River) (Saunders 2000:15), specification of the northern boundary is under dispute (Bushnell 1994:60; Hann 1987:2–4; Jones 1978:186; Saunders 2000:15), which results in part from recognition of the somewhat artificial nature of Guale as a distinct

Spanish province (Milanich 1999:41). Even more problematic is recognition that the Georgia coastal chiefdoms were certainly affected by epidemics during the early 16th century (Barcía 1951:153; Elvas 1907:173; Zubillaga 1946:418). Therefore, population size estimates derived from the early phases of missionization (ca. 1570) may significantly underestimate precontact population size. Timucua is a similarly abstract category suffering from poor boundary definition and reification issues. Apalachee, however, does not suffer these limitations to such an extent.

Baseline Estimates

Recognizing the lack of true precontact population estimates requires a simulated approach based on recorded village enumerations and general statements about settlement and village density. Estimates for Guale vary widely. One of the earliest sources indicated that 22 villages were present within the province (Lanning 1935:22). Pedro Menéndez recorded 40 Guale towns in 1565 (Lanning 1935:12), a figure that coincides well with Jones's enumeration listing 42 villages (Jones 1978:205–209). Milanich (1999:43) estimated 60 villages. Village size estimates also vary considerably. Milanich (1999) proposed 200–300 individuals per village. The Jesuit priest Father Sedeño provided two estimates of village density for the year 1570: 40 total individuals on average per village and 30–40 males on average per village (Jones 1978:190). Assuming that male population size refers to adult males only, and further assuming equal sex ratios and an equal number of children, then the population may have been around 100–200 individuals per village. These numbers result in a mid-16th-century population size of approximately 8,000 to 12,000 individuals for Guale. Milanich, using a more broadly inclusive definition encompassing a greater portion of the Georgia and South Carolina coasts, estimated a precontact population size of 31,000 individuals based on similarities in settlement density with eastern Timucua chiefdoms (Milanich 1999:45).

For Timucua, precontact population size estimates vary widely from a maximum of 670,000-plus by Dobyns and Swagerty (1980) to much more modest estimates in the tens of thousands (Deagan 1978; Milanich 1978). Worth reconstructed an average local chiefdom size of 1,500 individuals based on an expected 300 person per village density and an average of five communities comprising each chiefdom (Worth 1998b:4). For Yustaga, the population size was estimated at 6,000–12,000 individuals, essentially similar to that based on the early contact period warrior count discussed above (Milanich 1978). Based on differential settlement densities and recorded differences in alliance composition, Worth estimated a total interior Timucua population size of 27,000 individuals, and he concluded, "It must be remembered that the figures used as baseline populations in this chapter represent the interior Timucuans following more than half a century of sporadic European contact. The original populations of the prehistoric Timucuan chiefdoms of Florida's interior might well have exceeded 50,000 . . . suggesting that a 1492 Timucua-speaking population of 150,000 is probably a sound estimate" (Worth 1998b:8).

Therefore, Worth concedes that the estimates of Milanich (1996) of 150,000 individuals and Milanich (1999) of perhaps 200,000 individuals is a reasonable precontact population size figure for all Timucuan-speaking chiefdoms in Florida and Georgia. This number has received general consensus among historians (Hann 1996; Milanich 1996, 1999; Worth 1998b). Milanich (1999) has proposed similar population densities for the Timucua and Guale chiefdoms, approximately 10 people per square mile. Settlement density differences between Apalachee and their eastern neighbors suggest a precontact population density of approximately 40 people per square mile for Apalachee (Milanich 1999:50), which results in a precontact population size of 50,000 individuals within this small, circumscribed province. Hann (1988) cites several period enumerations of between 25,000 and 30,000 Apalachee at the time of first contact.

Rates of Reduction

In describing rates of population reduction, the most controversial decision lies in the initial population size estimate, which determines the percentage reduction for all succeeding estimates. For Guale, I adopt the low estimate of 8,000 individuals based on village densities and sizes as a reasonable average value. Rejection of Milanich's (1999) estimate of 31,000 individuals results from the work of both Griffin and colleagues (Griffin 1993; Griffin et al. 2001) and Stojanowski (2004), which suggest that the Irene Mound chiefdom was biologically affiliated with interior Creek/Lamar populations and *not* with the coastal Guale, despite the proximate nature of these polities. In other words, the estimate of 8,000 individuals applies only to the three subchiefdoms identified by Jones (1978) and does not include the Orista or Escamaçu-Ahoya chiefdoms. Density estimates for all Timucua chiefdoms seem to have reached a consensus of approximately 150,000 individuals (Hann 1996; Milanich 1996, 1999; Worth 1998b). For Apalachee, Hann's (1988) estimate of 30,000 individuals seems reasonable and is little contested and, as with the Guale, is on the low end of the ranges discussed, thus making the rate of reduction estimates very conservative.

Readers interested in the gory details of contact period population size estimation should consult the excellent discussions of Worth (1995, 1998b:1–26), Hann (1988:160–180, 1996:257–268), Deagan (1978), and Milanich (1978) for specific examples based on paleographic sources. Henige's (1998) treatise is of a more general yet critical nature. I am only interested in the proportional reduction in population size for each province, as summarized in Figure 3.1. This figure, constructed from consensus mission period enumeration data, graphically demonstrates the stark reality of the mission experience.

For Timucua there was a 67 percent reduction in population size by the time of concerted mission expansion (ca. 1600); the nadir was reached sometime between 1600 and 1650, after which minor fluctuations likely represent the effects of the Spanish policies of reducción and congregación (Worth 1998b). The total losses were over 99 percent. For Guale the data present a similarly bleak picture. Population size had decreased 85 percent by the first decade of the 17th century, a larger proportional reduction in comparison to Timucua, but it declined at a slower pace afterward, requiring the remainder of the 17th century to reach the nadir at 99 percent reduction. Although the late-17th-century estimates were adjusted for congregación policies, for example, the commingling of Guale and Mocama after 1660 and the post-1650 influx of Yamassee populations (Worth 1995), there is some year-to-year fluctuation that may suggest the nadir was achieved earlier than recognized. The eastern Timucua, western Timucua (with the exception of the Yustaga), and Guale populations were rapidly decimated by population size reductions, in stark contrast to the Apalachee who were, surprisingly, the most nucleated and sedentary mission community (Hann 1988). By 1600 Apalachee had experienced a 17 percent reduction in population size and reached a nadir of 73 percent by the first decade of the 18th century. Differences in rates of demographic collapse between the mission provinces were, therefore, very marked. This is consistent with other biological indicators (Larsen 2001). While both Guale and Timucua suffered rapid and catastrophic losses by the turn of the 17th century, Apalachee losses were never as severe, until the end of the mission period.

Spanish Responses to Demographic Collapse

Both the colonial and indigenous communities reacted and adjusted to the ever changing social landscape of 17th-century *La Florida*. For the Spanish, primary interest was the maintenance of their failing colony. The countenance of sustainability and a functional system of transport and communication were essential for mutual defense and for maintaining the limited economic potential the colony offered. Invariably, Spanish policy can be viewed as a progressive process of reorganizing the indigenous communities, often with little regard for the social repercussions of their actions. That is, the Spanish were involved in a numbers game and were generally insensitive to ethnic and cultural diversity. While population aggregation was, in some ways, beneficial for the indigenous populations as well (contemporary commentary indicates that small local population sizes placed further stress on local populations through workload requirements), this process was principally of Spanish design and planning and was of particularly significant consequence after 1650. Native Americans, how-

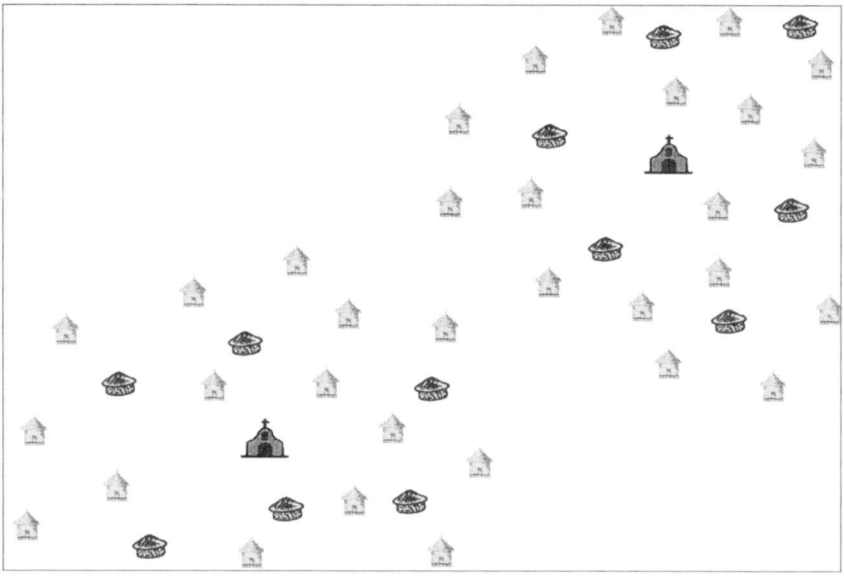

Figure 3.2. Two precontact chiefdoms are portrayed in this figure. The site of the paramount chief is represented by the mission *doctrina*, satellites under the jurisdiction of the paramount are represented by round aboriginal buildings, and individual households by small rectangular structures. Individual households obviously were more numerous than can be presented in this diagram; however, the hierarchical structure is apparent.

ever, also responded to the changing context of their lives, often in ways more ephemeral to the historian.

Spanish response to impending demographic collapse in the coastal and interior Timucua populations can be characterized as a staged process of exponentially aggressive population aggregation, relocation, and replacement (Worth 1998b). I stress at the outset that the changes experienced and the policies adopted refer most specifically to the Guale, eastern Timucua, and, to a lesser extent, western Timucua populations. As discussed, Apalachee never reached the population nadir typical of the coastal groups, and therefore it was spared some of the more drastic Spanish reorganization efforts. However, this is not meant to imply that the Apalachee were not affected by circumstances in the eastern provinces. Submission to Spanish sovereignty, as well as the developing conflict over European New World supremacy, guaranteed that the effects of colonization would be widespread and ubiquitous. This fact, in part, influences my decision to discuss Timucua at all, given the dearth of bioarchaeological resources attributed to this language group.

In Figure 3.2, I present a typical mission community structure with a central mission doctrina established at the seat of a regional chiefdom. The doctrina serves approximately 5–10 satellite villages subsumed within the jurisdiction of the regional *cacique*, with local chiefs organized in a feudal system of hierarchical labor and military tribute (Deagan 1978; Hann 1996; Milanich 1978, 1996, 1999; Worth 1998a, b). It is well established that the Franciscans opportunistically worked through the regional political structure to effect conversion of the subordinates. Although the size of the population served by the doctrina would vary considerably, and the number of local satellite villages would equally differ, this figure can be considered a general model of political structure that existed among the Guale, eastern Timucua, and western Timucua chiefdoms (Deagan 1978; Hann 1996; Jones 1978; Milanich 1978, 1996) and, to a lesser extent, the Apalachee (Hann 1988). Settlement densities are represented by icon size with individual households, local aggregate villages, and seats of regional paramounts also represented. Regional paramounts would in turn compete with each other or form alliances for mutual defense, such as that which existed among the Utina, Yustaga, and Potano to counteract the large and powerful Apalachee chiefdom (Hann 1988, 1996; Milanich 1996, 1999).

Figure 3.3 presents the initial stages of aggregation. Although we do not currently have the archaeological resolution to detail specific rates and processes of local aggregation, we do know that the process of consolidation to the doctrina (Stage 1) occurred very early in the mission period and was a deliberate design of the Franciscans (Geiger 1937:143; Worth 1998b:28–29). Social and political upheaval and the dissolution of the precontact political structure must have been a dramatic development, which, when coupled with the impending demographic catastrophe, placed immediate stress on the missionized tribes. Therefore, the most immediate and obvious feature of the earliest stages of missionization was the loss of population numbers. Disease, of course, was the primary cause; however, fugitivism was also problematic and is a known strategy of avoidance beginning in the late 16th century (Jones 1978; Lanning 1935). Ultimately, slave raiding, reduced fertility, and deaths due to labor draft hardship would increase rates of reduction, which were particularly acute after 1650 (Worth 1998b, and see Larsen 2001) and would eventually place the onus of "civilization" squarely on the shoulders of the Apalachee.

The rate of early depopulation quickly stressed the Franciscan infrastructure. Friars charged to serve and convert the entire jurisdiction were challenged to meet the needs of the increasingly widely distributed and dispersed populations. Administration of last rites, participation in Christian lifestyles, and easing the burden of living in a small and dysfunctional community served as the political "spin" selling the need for local aggregation; the colonial government found nothing adverse in these Franciscan requests: "if the smaller towns, i.e., those outside the places governed by a head cacique which had as few as three, four, or five houses, were to be effectively converted, the king should command them to form one settlement with the larger town" (Father

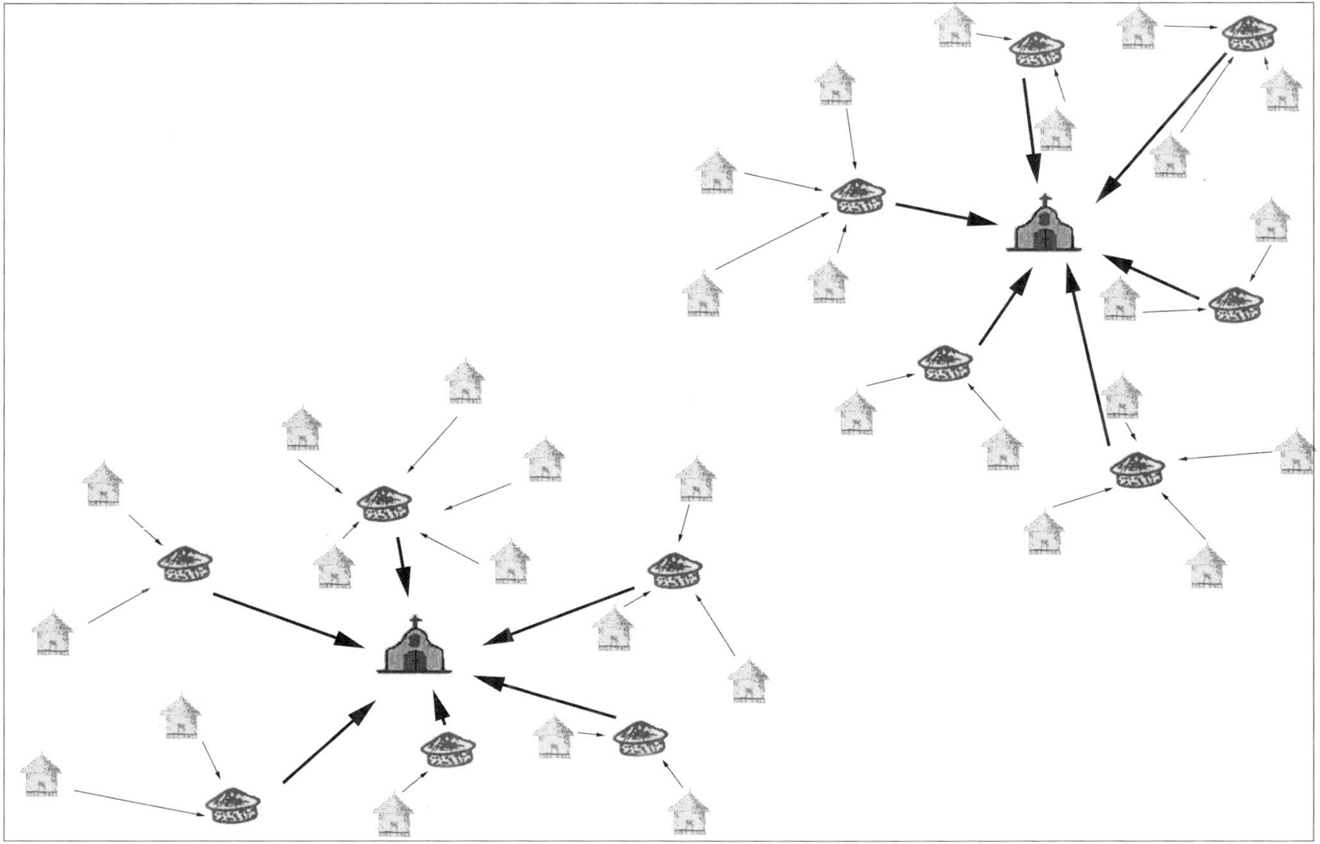

Figure 3.3. Stage 1 *congregación,* that is, the consolidation of households to the local village level and the movement of populations from the local villages to the site of the primary *doctrina.* Because doctrina structure paralleled precontact indigenous political organization, it is unlikely that such aggregation was significant from an evolutionary perspective because the individual households and local villages were likely closely related to each other during the precontact period (modified from Larsen 2001: Fig. 2.1).

Bermejo, Letter to the Crown, in Geiger 1937:143). The years 1600 to approximately 1630 can therefore be viewed as a period of intensive local aggregation as diffuse and semisedentary populations fully accepted the Christian ideal of settled village life (Deagan 1978:113; Worth 1998b).

Stage 1 aggregation was of dubious evolutionary significance. Populations once united under a common political entity were now congregated into an administrative unit of Spanish colonial structure. Period documents support this proposition:

> We request that Your Majesty would be served to command that, whenever these necessities occur, as long as the governors are advised by the prelate, these disordered [settlements] should be drawn together, since there is not one inconvenience, *through those that have to join together not being from different families or languages, but rather friends of friends, brothers of brothers, and relatives of relatives beforehand* [emphasis added]. It is an important and necessary thing for them, since the reason for which so many Indians die is customarily that being few, they cannot help each other in their labors. (Worth, 1998b:28–30)

Continued depopulation owing to disease, fugitivism, slave raiding, and repartimiento-related deaths during the early to middle 17th century required more concerted Spanish intervention. While local aggregation was an obvious and inescapable first approach, it was ultimately unsuccessful. The next phase involved multiple corrective attempts that varied in their intensity by region. Concerted effort at *cimarrone* (fugitive) recovery was one measure adopted by the Spanish (at least 12 expeditions to recover Timucua fugitives were launched between 1620 and 1655 (Worth 1998b:23); however, bringing the unwilling back to God was more difficult than rewarding the loyalists (and opportunists) for their relocation and cooperation. Figure 3.4 depicts the widespread and well-documented practice of congregación, the aggregation of formerly distinct doctrinas into a single, larger functional doctrina (Stage 2) (Worth 1998b). The process of doctrina consolidation was of direct Spanish design and was often intended to maintain functional populations at critical places in the communication and travel system. Maintaining ferry crossings for major rivers (Worth 1998b), restructuring for mutual defense of the missions (Worth 1995, 1998b), and maintaining way stations (approximately one day's journey by foot) along the mission road into the interior were common reasons for congregación. Preexisting linguistic, cultural, or political structures were certainly not of concern to the Spanish.

Worth's landmark historical analysis of the Guale, Mocama, and St. Johns River valley missions provides detailed accounts of the doctrina consolidation process. The Mocama missions declined from four in 1655 to two in 1670

Figure 3.4. Stage 2 *congregación,* that is *doctrina* consolidation within organizational provinces (Apalachee, Guale, Timucua). Small arrows represent previous aggregation of local villages at the doctrina (Type 1 congregación). Large arrows represent the consolidation of distinct doctrinas (Type 2 congregación). Because doctrinas were initially associated with political epicenters, the consolidation of individual doctrinas into large, aggregate mission complexes may have resulted in the aggregation of distinct political and biological units. Consolidations labeled 'A' represent aggregation of doctrinas within a political and linguistic region (within Timucua-speaking populations); those labeled 'B' represent consolidation of groups that speak similar languages and were enumerated by the Spanish under a single label (Timucua) but that likely represented distinct precontact chiefdoms. The arrow labeled 'C' represents the hypothetical movement of an Apalachee mission to Yustaga. Such aggregation is considered Type 1 *reducción* in this model (relocation of other Christian populations) and would provide the largest opportunity for increased genetic variability (modified from Larsen 2001: Fig. 2.1).

(Worth 1995). The Guale missions experienced similar doctrina aggregation, declining from six distinct doctrinas in 1666 to three in 1680, one of which itself represented the remnants of four distinct missions that had existed just 14 years prior (Worth 1995). Considering that each doctrina may have served the population of 10 to 20 local villages, these descendants represented a small fraction of the ancestral population. More interesting, Worth presents the impetus for this restructuring, the slave raiding by English-allied indigenous populations beginning as early as 1661; eventually, direct English assaults on the coastal missions lead to Guale's abandonment and relocation in the middle of Mocama, an area formerly inhabited by Timucua speakers (Worth 1995).

The effects of congregación in terms of settlement distribution were dramatic. Eventually, after mid-century (and a particularly devastating uprising in the Timucua interior), all interior Timucua settlements were relocated near the mission road (Figure 3.5), documented archaeologically by a change in site distribution from a density model of circular orientation to one that is more linear representing the alignment of the population along the *camino real* (royal road) (Johnson 1991). Doctrina consolidation in the coastal provinces also has a distinctly southward orientation (Figure 3.5) as the Guale are moved to the barrier islands, then eventually ever southward toward St. Augustine (Worth 1995), and the eastern Timucua are similarly aggregated near St. Augustine or at important locations in the St. Johns River drainage (Worth 1998b). The process of congregación does not seem to have occurred to such an extent in Apalachee as mission lists in 1647, 1657, and 1689 indicate stability in the number of villages (Hann 1986b:373, 1988:100).

When congregación failed to effect demographic stability for key mission locations, a reducción policy was enacted. Reducción entailed creating entirely new villages or repopulating a vacant doctrina with individuals from nonlocal populations. I consider two important distinctions: reducción of populations already missionized, leading to a restructuring of *Christian* populations (Type 1), and reducción of pagan groups external to, and sometimes hostile to, the Spanish colony (Type 2). While the conflation of Guale and Mocama during the middle 17th century might be considered Type 1 reducción, this process was simply the result of population backflow as the cultural landscape around St. Augustine (that is, the coast and St. Johns River region, not the interior Timucua and Apalachee) was transformed prefacing the inevitable final retreat to St. Augustine proper. More pointed examples of Type 1 reducción occurred for the interior Potano and Utina districts after the 1656 Timucua rebellion when Yustaga were implored to repopulate missions in Utina, Potano, and some of the eastern districts (Hann 1996:225; Worth 1998b).

Historical examples of Christian population reorganization (Type 1) are additionally supported by archaeological evidence for changes in the distribution of ceramic types during the 17th century (Deagan 1990a; Hann 1986b:379,

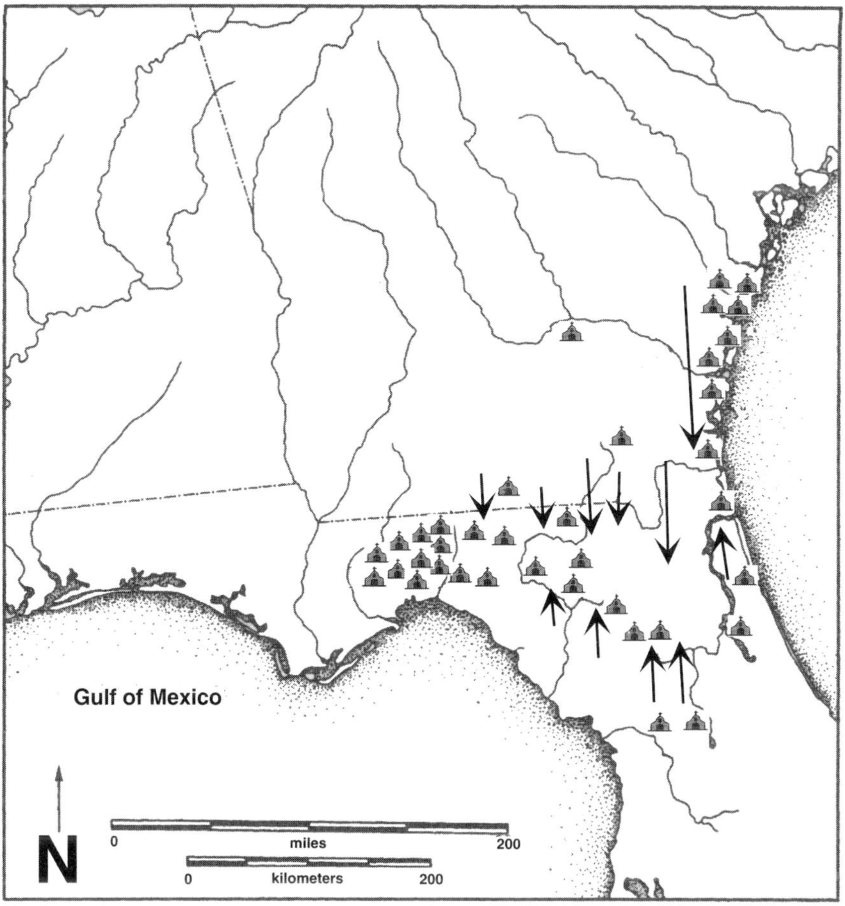

Figure 3.5. This figure demonstrates the transformation of the Timucua interior and Guale and Timucua coast after demographic collapse had commenced. There is a general shift of populations toward the mission road as Spain attempted to maintain the viability of its colony and a chain of communication linking St. Augustine with the Apalachee interior (modified from Larsen 2001: Fig. 2.1).

383, 384, 1996:232; Milanich 1978:70, 1996; Saunders 2000; Weisman 1992: 166; Worth 1992:171–182), which could represent either the replacement of peoples (migration) or the dissolution of precontact barriers to materials exchange. The general picture demonstrated by the ceramic data is one of contraction toward St. Augustine, with the interior Timucua region representing the gravity epicenter of backflow.

Archaeologists have documented a gradual replacement of precontact ceramic types in the Timucua interior (Suwannee Valley and Alachua types) with Leon-Jefferson types indigenous to Apalachee and Yustaga (Hann 1996:85;

Worth 1992). Milanich (1972:58) investigated ceramic distributions at the Zetrouer and Fox Pond sites, both associated with the interior Potano. He documented an increase in Leon-Jefferson wares and also a high frequency of San Marcos wares, associated during the precontact period with the Guale or Yamassee. Leon-Jefferson types, previously restricted in distribution to the Apalachee and Yustaga region, have been found in contact period sites as far east as St. Augustine (King 1984). This transformation in ceramic technology provides additional evidence for the west-to-east movement of people along the camino real during the second half of the 17th century and the north-to-south movement of populations along the Atlantic coast.

Among the Mocama and eastern Timucua groups of the northern Florida Atlantic coast, San Marcos (Guale) types also gradually replace indigenous St. Johns varieties at a number of mission and nonmission locations, again representing a southern contraction along the Georgia coast toward St. Augustine. Deagan has summarized these movements:

> The full acceptance of Guale cultural elements as replacements of eastern Timucua elements can be seen archaeologically in the replacement of St. Johns ceramics by Guale ceramics during the seventeenth century, a process which was virtually complete by 1700. Population movements are archeologically indicated by such sites as San Juan del Puerto, which yields evidence of patterning traits typical of the Tacatacuru, and by Wrights Landing, which indicates a sudden, non-Timucua population increase in the 17th century. (Deagan 1978:115)

While it is clear that each province was associated with a specific ceramic assemblage before contact (Saunders 2000) and that these distributions became more homogenous during the 17th century, historians have differed in their interpretation of the significance. Weisman (1992), while recognizing the same trend of replacement in the Florida interior, proposed a direct immigration of Creeks from the Lamar heartland in central Georgia. He cites evidence for population size increases early in the 17th century as support for this hypothesis (Weisman 1992:166). Hann (1996:234) explicitly denies any historical proof of actual population replacement. In particular, the continued use of Timucua and Apalachee interpreters in their respective provinces throughout the 17th century contradicts a hypothesis of complete population replacement by Creeks from central Georgia (Hann 1996:234). However, within the mission provinces, there is a clear trend of, at the very least, relaxed boundaries for the exchange of material culture, coincident with historical documentation of excessive migration and relocation of Christian populations during the 17th century.

Reducción of pagan populations (Type 2) is also documented in the historical record (Hann 1988; Worth 1998b). Often both forms of reducción were used, resulting in a veritable kaleidoscope of cultural and linguistic groups. The example of mission San Diego de Helaca is particularly salient. This mission first appears on the 1627 mission list and was created by Governor Rojas y Borja for the express purpose of maintaining the ferry across the St. Johns River. Within 13 years of its establishment, individuals from three distinct districts had populated the mission, the Acuera, the Agua Dulce, and the Mayaca (a nonlocal group completely unrelated to the Timucua) (Worth 1998b:32). After the Mayaca exodus, the government tried, ultimately unsuccessfully, to force the Chisca, a nomadic bellicose group who were actively raiding the Apalachee (Hann 1988:187), to settle near mission San Diego (Worth 1998b). In the short span of approximately 25 years, four populations, speaking possibly four different languages, were enlisted to settle this particular location. Such variability was typical of the St. Johns River drainage communities during the 17th century.

This policy of tolerance represented a general attitude adopted by the Spanish and apparently by the indigenous communities toward outside groups that previously may not have been allowed. Although intensive aggregation may not have occurred in the Apalachee region, there is some evidence to suggest that the peace initiated by the Spanish resulted in new opportunities for mate exchange within Apalachee province. Direct historical proof of resettlement is surprisingly rare, although Hann (1988) mentions a few instances. For example, 300 Tocobaga (from near Tampa Bay) had settled at Wacissa near San Lorenzo de Ivitachuco, 20 of whom were buried at Ivitachuco, apparently becoming Christians on their deathbeds (Hann 1988:165). Likewise, two non-Christian groups (the Chine and Chacato) were allowed to settle within the jurisdiction of San Luis, an event that required the petitioning of the mission's leadership and the construction of a formal contract outlining the immigrants' rights (Hann 1988:102). The Yamassee also make an appearance in Apalachee, but, unlike the Chacato, they were physically living at the San Luis village site (Hann 1988:103). Such in-migration is frequently used to explain increases in late-17th-century village sizes despite documented epidemics (Hann 1988:173–174). A 1675 enumeration indicates 649 non-Apalachee living in the province (Hann 1988), which represented 8–10 percent of the total population.

Similar examples of Type 2 reducción have been recorded for the Guale, with the appearance of 300-plus Yamassee in their ranks (Worth 1995), for the Mocama, also involving the Yamassee (Worth 1995), and for several St. Johns River populations, the Mayaca and Chisca (Worth 1998b). While there is little historical evidence for interbreeding between these groups, the shift in political orientation and the close contact would have provided new opportunities for mate exchange with previously biologically distinct populations. Gene flow seems a certainty.

Native Responses to Colonization

While the preceding discussion is well documented, European hegemonic policy was not the sole cause of evolutionary transformation in the mission populations. And while history is generally more silent with respect to Native American solutions to the demographic problem, circumstantial evidence suggests several proactive measures were adopted in the wake of demographic collapse. The preceding section clearly indicates that the mission period witnessed substantial new opportunities for admixture between culturally distinct populations. In addition, available data suggest these opportunities may have been acted upon, perhaps out of necessity.

Documented fugitivism was a potentially two-pronged process: out-migration and fugitivism. Individuals who truly left the Spanish sphere of influence in exchange for a less regimented life in the interior regions of Georgia, Alabama, and south and central Florida may have brought their families with them. Although I have seen no mention of this, it seems likely that kin-structured out-migration was a common strategy of the truly disenfranchised. However, fugitivism also became an acutely epidemic problem for native males, particularly after mid-century.[4] Repartimiento labor drafts required that males be away from their families for extended periods of time, and the problems associated with this form of transiency did not go unreported. For example, the 1682 Synod of the Diocese of Santiago de Cuba, Jamaica, Habana, and Florida stated "that the married Indians learning the doctrine in Florida do [*sic*] not be kept a long time in the City of St. Augustine far away from their wives" (McEwan 2001:635), a direct reference to issues related to the repartimiento quotas.

Disdain for the labor tribute was ubiquitous. Native elites generally had little problem allocating the labor of their subordinates in exchange for the political legitimization they received. As population sizes dwindled, however, the Spanish made little distinction between the hauling, building, and work capabilities of elites and commoners. According to Worth (1998b), transgressions related to repartimiento expectations were one of the primary reasons for significant resentment and rebellion. A tried-and-true method to avoid excessive labor demands was intentional absence from natal villages. The temptation to flee was further enhanced by developing economic opportunities for males in the interior ranching industry.

Attempts at fugitive recovery were staged in an effort to locate absent males (Worth, 1998b). One such example is the recovery of an Indian named Lorenzo of San Diego de Helaca who was living in the village of San Antonio de Enacape in the Ibinuiti district. The leader of the military expedition was ordered to "bring a married Indian called Lorenzo . . . so that he makes a life with his wife and serves in it" (Worth 1998b:32). The severity of the problem required gubernatorial intervention (Hann 1986a:102, 103).

For Apalachee, problems with fugitivism clearly increase after Spanish demands for repartimiento labor increase, no doubt in response to the devastation in the eastern coastal provinces. In 1694, the chief of Capoli requested that four husbands be returned to his village because they were "missed by the village as well as by their wives" (Hann 1988:171). According to Bushnell (1979:5), nearly half of the men were absent from a single Apalachee village in a single year because of repartimiento requirements. The long-term absence of males may, therefore, have forced Apalachee women to choose mates from nontraditional, and socially unsanctioned, sources.

It is difficult to estimate what effect male transiency may have had on indigenous mating structure. There is, however, one documented example of admixture between a Christian and a pagan individual. In reference to the 1647 Apalachee revolt, Hann writes, "One of the ring leaders, Juan de Diosale, was referred to as a great tascaya, the son of a Chisca, greatly feared because he was part Chisca and had a following among the Chisca, some of whom were always to be found at his house" (Hann 1988:184). This is a particularly salient example because the Chisca were a perennial problem for the Spanish; they instigated the 1647 rebellion in Apalachee (Hann 1988:184), but at the same time they were the target of a 1685 attack launched by the cacique (chief) of San Luis against a Chisca village. Such confusion highlights the complicated relationship among linguistic, cultural, and political factors and patterns of genetic admixture in polyethnic communities.

Therefore, the male demographic problem, combined with the migration occurring between mission groups (Type 1 reducción) and the reducción of nonlocal pagan groups (Type 2), provided ready opportunity for mate exchange previously unavailable because of sociopolitical boundaries. The evolutionary effects of this behavior are difficult to estimate with certainty. However, ethnohistoric data from the precontact period suggest the cultural and linguistic boundaries documented by early Spanish and French explorers were biologically significant. Therefore, this process of accommodation is expected to have had evolutionary repercussions for the Christian populations (see below). Regardless of the extent and nature of indigenous population relationships, opportunities for mate exchange with nonaboriginal populations would certainly have affected patterns of genetic diversity.

Evidence for European and African Admixture

Deagan (1973, 1983) and McEwan (1991b) have synthesized much of the literature on New World *mestizaje* (Spanish-indigenous admixture), particularly pertaining to 16th- and 17th-century *La Florida.* Data on the incidence of mestizaje is lacking for Apalachee (Hann 1988:237), but Deagan (1990a) notes that contemporaneous records indicate that Guale females comprised a significant portion of the wives of the Spanish.

Ethnohistoric accounts of European admixture predate the establishment of missions. Barcía recorded two instances of Frenchmen taking Indian brides in the mid-16th century; one involved a colleague of Laudonnière (Barcía 1951:51) and the other a member of Ribault's party (Barcía 1951:113). He also relates a comparatively rare instance of indigenous admixture involving a European female. After a shipwreck in the Straits of Florida, he records, "Three or four of the women who had survived the disaster were still living in the country of Carlos, were married, and had borne children" (Barcía 1951:51). Likewise, Ranjel's testimony is riddled with references to women being provided to de Soto and his men as a gesture of peace (Ranjel in Bourne 1904: 113, 115, 117, 122, 144, 145). The exact nature of this interaction is nonspecific but is likely sexual in nature.

Intermarriage with the Spanish began almost immediately after the founding of St. Augustine. One of the earliest recorded examples involved Pedro Menéndez himself, who was married to the sister of Carlos, the Calusa chief (Lyon 1976:148). However, as Menéndez already had a Spanish wife, this marriage was likely more of a political gesture. Dona María Melendez, the late-16th-century *cacica* of the Nombre de Dios mission, was married to a Spanish officer stationed at St. Augustine (Deagan 1973:58, 1985:304). It is important to note in both examples the distinction between marriage as a social contract and mating as a biological process. Such interaction was of intentional design: "While the acquiescence to the Spaniards benefited native leaders in the form of political recognition, titles, and exotic gifts, it often had a negative effect on village and family life. Native social structure was further impacted by *mestizaje,* or the intermarriage and interbreeding of Spaniards with natives, which Spanish authorities encouraged as a means of 'civilizing' and effecting religious conversion among the native populations" (McEwan, 2001:635).

Similar opinions predate the *La Florida* colony. In 1503, Queen Isabela stated it would be beneficial if "some Christians marry some Indian women and some Christian women marry some Indian men, so that both parties can communicate and teach each other, and the Indians become men and women of reason" (Morner 1967:26).

For Spanish men in the province (many of whom were soldiers), mestizaje seems to have been a necessity. Spanish female brides were in short supply, representing a minority of immigrants to the colonies (Boyd-Bowman 1968), which may relate to the apparent difficulty involved in securing the appropriate permits for single females (McEwan 2001:34). Of the 1,200-member crew of Menéndez's expedition, there were only 100 married couples (Dunkle 1958:4);

the remainder was single males. In 1578, only six additional women colonists were requested of the Crown (Dunkle 1958:4), which Deagan interprets as evidence that the Spaniards were finding mates among the Guale (Deagan 1973:57). A 1598 proclamation issued at the time of Governor Canzo's tenure as administrative ruler of *La Florida* declared, "The king informed the Casa de *Contratación* that with the newly appointed governor he was sending to Florida a secular priest, to serve as chaplain of the presidio, twenty-four soldiers, and ten or twelve women of good reputation as prospective wives for the soldiers of the fort" (Geiger, 1937:71–72). The dearth of Spanish females in this request suggests that the local Indian population provided a sufficient supply of brides for the Spanish military. Hann's analysis of enumeration data provides additional highly circumstantial evidence for mestizaje. For example, the 1604 census at Nombre de Dios included 20 Spaniards, which suggests, according to Hann, that they had indigenous wives (Hann 1996:164). Likewise, among the western Timucua, Hann concludes that the awareness of the soldiers of the impending 1656 rebellion resulted from their marriage to, and close contact with, Timucua females (Hann 1996:203). Pre-18th-century mestizaje is perhaps best documented at mission San Luis in Apalachee Province where it is known that some soldiers were married to Apalachee women (Boyd et al. 1951:24–26), as well as evidence for "Hispanic" households derived from the distribution of material culture in relationship to ethnic enclaves (McEwan 1991a, b, 2001). Additional evidence for Apalachee derives from the repeated proscriptions against concubinage levied by government officials: "The emigrés to Mobile made an obscure reference to this phenomenon [concubinage] as one of their reasons for wishing to stay among the French, remarking that at San Luis they had not been masters of their wives and noting that since they had been living among the French it had not been a problem" (Hann 1988:170). Although intermarriage between the Spanish and local Indian populations certainly occurred, it should be noted that, according to Deagan, a "Hispanic" population did not exist in *La Florida* until after the mission system had collapsed and populations were aggregated at St. Augustine:

> In Florida on the other hand, the mission system as the sole medium of contact, and the presence of a single settlement throughout the seventeenth century, combined to inhibit widespread miscegenation. This did not mean, however, that there were no Indian-Spanish unions with mextizo offspring. Archaeological and ethnohistorical evidence do indicate that this occurred frequently in colonial St. Augustine, although on a much more limited scale than in Mexico. The only intensive side-by-side living came after the relocation of the Florida Indian population in St. Augustine around 1700. It was only within this situation in St. Augustine

> that *mestizaje* occurred to a significant degree, and this was brought to a premature end with the removal of the Spanish from Florida in 1763. (Deagan 1973:56)

Baptismal and marriage data from the St. Augustine parish are informative. Between 1735 and 1750, 18 percent of the marriages were between Africans and mulattos, 71 percent were between Spaniards, and 11 percent were between *mestizos* and Indians, Spaniards and Indians, or Spaniards and mestizos (Deagan 1973:59). For the same time interval, 27 percent of the births were classified as African or mulatto, 70 percent were European, and 3 percent were mixed Indian-Spanish (Deagan 1973:59). By 1760, 8–10 percent of the entire population of Spanish Florida was classified as mestizo (Deagan 1973:60). Interestingly, there were only two Indian-African marriages and no marriages of male Indians with female Europeans (Deagan 1973:60). All mestizo marriages involved Guale or Yamassee; only two Apalachee were included in this enumeration (Deagan 1973:60).

African slaves were also a feature of the early colonization efforts in Florida (Barcía 1951:70; Bolton 1921:141; Deagan 1985:295). In 1603, the African population of St. Augustine was listed as 32, five of whom were women (Geiger 1937:168), and in 1604 the colonists requested a dozen additional slaves and three or four African females (Dunkle 1958:5). By the end of the Spanish occupation of Florida in the mid-18th century the African population had grown considerably. There were 87 free Africans—31 men, 34 women, and 22 children (Siebert 1940:146)—and 303 slaves, including four men and five women designated as admixed (Siebert 1940:147).

Direct evidence for African-indigenous admixture is also extremely rare, but proscriptions against the practice suggest it too was common where opportunity existed (Boxer 1975). Geiger (1937) relates one story in relationship to seven slaves who had escaped from St. Augustine in 1603. Governor Ybarra recaptured five of these individuals; however, two remained among the Ais where they took Indian brides (Geiger 1937:177). According to Morner, mating between the Indian and African populations was widespread throughout Spanish Florida, despite the objections of the Spanish government (Hanke 1964:31; Morner 1967:40; Wright 1981:44). For St. Augustine, Deagan comments, "St. Augustine parish records reveal that marriages between both free and slave blacks and Indian men and women took place in Florida with some regularity from the mid-17th century onward. Nineteen of these marriages occurred between 1675 and 1750, accounting for 12.3 percent of all racially mixed marriages during that time" (Deagan 1985:306). Additional evidence for Spanish concern with African-indigenous admixture comes from Francisco Pareja's Timucua confessionary. The pointed question "Have you some [black female] slave or servant as your mistress?" (Milanich and Sturtevant 1972:34) is highly suggestive. It is, however, ultimately difficult to differentiate whether this question was proactive or reactionary in nature.

Precontact Population Structure

The mission period can be aptly summarized as a period of population size decline, population restructuring, aggregation, and in-migration, with concomitant shifts in social behavior owing to both necessity and new opportunities for genetic admixture with multiple linguistic and ethnic groups. The question remains how significant such turmoil would have been from a population genetic perspective. While it is clear that African or European admixture would certainly have increased population genetic variability, it is less certain how significant population restructuring would have been in terms of the commingling of distinct indigenous mating networks.

Ascertainment of expected effects requires baseline information on the manner in which populations were defined during the precontact period. Given recent debates over the fluidity of indigenous social barriers (Moore 1994a, b; Quinn 1993; Terrell et al. 1997) and documented historical examples of intertribal integration (Albers 1993; Moore 1987, 1994a, b; Owen 1965; Sharrock 1974), it is extremely difficult, if not impossible, to generate statements about the *absence* of integration. The "absence of evidence" axiom certainly applies in this context. Nevertheless, certain kinds of ethnohistoric data may be useful in generating statements about potential culture contacts. We might be able to say populations X and Y were more likely biologically integrated than X and Z, based on sociopolitical observations of contemporaneous observers. In addition, documentation of intermarriage or mate exchange between different ethnic units does provide evidence of the degree of mate exchange that may have characterized indigenous communities.

A suite of data was culled from ethnohistoric discussions in an attempt to establish population boundaries that existed at the time of initial European contact. In so doing, I attempt to predict at what stage the congregación process (Stage 1 and Stage 2) or the reducción process (Type 1 or Type 2) is expected to elicit an increase in population genetic variability. As outlined above, seven types of data were identified in the ethnohistoric literature dating to the earliest years of colonization: political structure, patterns of conflict and warfare, linguistic differences, sedentism and residential mobility, patterns of trade and exchange, marriage patterns and social organization, and evidence for cross ethnic marriage. It should be noted that, with rare exception, the inference available from such reconstructions is one of degree rather than of kind. Mating is more or less frequent rather than a dichotomous presence/absence condition.

Political Structure. Understanding the political structure at the time of contact is a crucial first step in estimating population boundaries. While political borders are by no means a complete barrier to gene flow, it seems reasonable to conclude that mating was more frequent within political territories than between them. One obvious exception is the arranged marriage designed to solidify newfound loyalties among competing polities. The most common references to political divisions are based on the use of specific names of villages or chiefs that indicate a clear identity division among native communities. The de Soto chroniclers are particularly astute recorders of such information (Bourne 1904; Elvas 1907), noting the presence of multiple polities by name as well as the distances that separated them. Of particular interest are notations related to the uninhabited buffer zones providing a physical barrier between competing and warring chiefdoms (DePratter 1991). Such buffer zones isolated the Apalachee from their neighbors (DePratter 1991; Hann 1988; Steinen and Ritson 1996) and the Guale from interior chiefdoms (Jones 1978). Macrolevel divisions are relatively easy to identify. More difficult to identify are microlevel distinctions, such as those purported for Guale (Jones 1978) and Apalachee (Hann 1988). At this time, too little is known about the significance of such divisions to generate mating predictions. In my opinion, however, micropolitical structure had little impact on patterns of gene flow within Spanish-recognized political units (Apalachee, Guale).

Warfare and Conflict. Evaluating patterns of warfare provides information on the degree of interaction that two populations may have experienced. Fortunately, the ethnohistoric record provides specific references to conflict that can be used to differentiate populations. The Apalachee, for example, were feared throughout Florida; they were known to have warred with the Yustaga, Utina, and Potano confederacies to their east and with the Apalachicola, Chine, and Chacato chiefdoms to their west (Steinen and Ritson 1996:113). Various references portray the pattern of warfare among Timucua populations, and the Guale are likewise distinguished by internecine warfare with the Mocama to their south and from the Orista to their north (Lanning 1935:38). Does inclusion within a polyethnic and polyglot confederacy indicate that gene flow was concomitant? This is difficult to verify with certainty but may be likely. Another difficulty with using conflict data to discern mating boundaries is the nature of indigenous warfare, which was often small-scale and which frequently resulted in the capture and incorporation of females into the chiefdom of the victor (DePratter 1991; Laudonnière 1975). Laudonnière's time among the coastal Timucua during the mid-16th century produced these observations: "The kings make great war among themselves, always by surprise attack. They kill every male enemy they can. They cut the skin off their heads to preserve the hair, and carry this back on their triumphant homeward journeys. They spare the enemy women and children, feed them, and retain them permanently among themselves" (Laudonnière, 1975:11). DePratter (1991:51, 52) indicates that the capture of women may have increased the agricultural productivity of the victor and provides an example from the 16th century on the manner in which captured females were divided among chiefdoms within an alliance. Garcilaso's account indicates that captives were commonly encountered during de Soto's march throughout the Southeast (Varner and Varner 1951:329). Therefore, although warfare may have precluded reification of mate exchange within prevailing systems of social organization, that two chiefdoms warred may not have been as genetically limiting as one might first assume. It seems likely, however, that submission within a confederacy was more homogenizing a force biologically than was warfare.

Linguistic Differences. There are few original documents that contain text of Apalachee, Timucua, or Guale language (for example, Milanich and Sturtevant, 1972). However, the limited information that can be marshaled suggests that three distinct and mutually unintelligible languages were spoken. Internal linguistic consistency is suggested for Guale and Apalachee (Hann 1988), while the Timucua chiefdoms spoke a dozen or so dialects with varying degrees of mutual intelligibility (Deagan 1978; Hann 1996; Milanich 1978, 1996, 1999). Language functions much like warfare in terms of biological patterning. Mating is possible, of course, between people who speak different languages, and Spanish identification of linguistic geography does not necessarily equate with reality when multilingualism may have been the norm (Moore 1994a, b). Period interpreter requests and travel logs reveal the degree of linguistic diversity and mutual intelligibility characterizing southeastern communities (Lanning 1935:11; Swanton 1922:11). Elvas recorded that by the time de Soto reached the Mississippi interior, he had brought with him a dozen or more interpreters who would be required to translate the local languages (Elvas in Bourne 1904: 146). As with warfare patterns, speaking a common language likely indicates higher probabilities of gene flow between populations, but linguistic diversity is not, by itself, a barrier to gene flow.

Sedentism and Residential Mobility. Movement equals opportunity. Polities that are densely settled and agricultural chiefdoms surrounded on all sides by enemies with whom conflict was incessant would have had less opportunity for long distance mate exchange owing to the political landscape and would have had less need for long distance mate exchange owing to the locally sufficient population base. Small mobile or semisedentary populations that ranged widely throughout the year would have greater opportunity for admixture with nonlocal populations. The moderate correlation between subsistence, population size, settlement density, and sedentism suggests that the Apalachee were limited in opportunities for long range genetic admixture. In addition, with a precontact population size estimated between 30,000 and 50,000 individuals (Hann 1988; Milanich 1999), the Apalachee may have had no need to seek mates in such a manner. Their lack of popularity in the regional political scheme assures that such behavior was uncommon.

The Apalachee are contrasted with the small, semisedentary populations of the eastern Timucua districts where population sizes were small (some numbering less than 1,000 individuals; Deagan 1978) and where the food quest was wider ranging. In such situations it seems unlikely that a semisedentary band was large enough to maintain any degree of endogamy, indicating that mate exchange between "ethnic" groups was common. If degrees of mobility and population sizes of cultural groups decrease as one moves east from Apalachee toward the Atlantic coast, then one may expect the effects of Spanish reorganization to be less biologically salient at lower levels of the aggregation process. For example, both Stage 1 and 2 congregación may have been minimally significant for Apalachee and for eastern Timucua groups even if the doctrinas consolidated were identified with distinct cultural groups. Incorporation of non-Christian populations (Type 2 reducción) may have been more significant for the Apalachee than for the eastern Timucua groups owing to the biological divergence associated with political isolation and sedentism.

Trade and Exchange. It is difficult to evaluate the significance of data on resource and materials exchange from a biological perspective. Certainly, trade interaction over wide geographic areas increases the probability of gene flow between populations. And, as with alleles, extensive trade networks can result in the exchange of material culture without actual contact between the involved parties, making inferences of the biological significance of trade and exchange more tenuous. It does, nonetheless, provide some information about potential cultural contacts and the size of the interaction sphere in which mating may have occurred.

Marriage Patterns and Social Organization. Information on social organization may provide evidence of the prescribed structure of mating, particularly as related to clan affiliations, patterns of endogamy or exogamy, and proscriptions relating to choice of mate. It is unlikely that such information will consider between-province relationships, which should not be construed as evidence for a lack of gene flow (Moore 1994a, b). However, information on marriage patterns and social organization can be useful for inferring processes internal to an ethnic or political division.

One example of this comes from the Apalachee ball game manuscript (Bushnell 1978b). According to Bushnell, when intervillage play commenced, village identification was based on the color of team uniforms which represented the totems of the dominant clan of that village (Bushnell 1978b:13). Given the dominance of clan exogamy in the Southeast (Hudson 1976), this fact suggests that villages were affiliated with major clans and that exogamy required individuals to marry beyond their local group. This implies that Apalachee villages were biologically integrated and therefore regionally homogenous. This kind of structure suggests that both Stage 1 and 2 congregación would have been a non-event in terms of Apalachee evolutionary history.

Direct Evidence for Intermarriage. One of the most valuable but least prevalent classes of information is direct evidence of intermarriage between individuals affiliated with different cultural or political groups. Such information exists for the Calusa (Goggin and Sturtevant 1964:189) and for the late-17th-century Apalachee (Hann 1988), but I have found no such evidence recorded in the immediate postcontact period documents. Such anecdotes are the "smoking gun" that biologically links different ethnic groups together. Although useful in a binary sense, such information tells us nothing about the actual rate of intermarriage, however.

Research Models

Two competing evolutionary mechanisms were operating in 17th-century *La Florida* communities. First, demographic collapse and associated allele fixation/loss were decreasing within-population genetic variability, perhaps differentially reflected by differences in population size and changes in population size through time. Second, aggregation, relocation, and gene flow between distinct allochthonous sociopolitical units and indigenous communities, and between African and European New World migrants and indigenous communities, was increasing population genetic variability.

Observations of genetic variability in a single lineage for two time periods can produce stasis, an increase, or a decrease in population genetic variance estimates. A temporal decrease in variability implicates genetic drift as the primary evolutionary mechanism and suggests a decrease in effective population size in the population. A temporal increase in variability implicates aggregation and gene flow as the primary evolutionary mechanisms affecting phenotypic variability and suggests a change in the migration history, mating patterns, or boundary definitions for that population. Stasis suggests that neither population size nor gene flow patterns changed during the survey interval.

The interplay of ethnohistory and human biology creates a dynamic predictive model for generating evolutionary research hypotheses in historic-era populations. Epidemic disease, frontier slave raiding, fugitivism, and deaths due to mission-related hardships all contributed to the demise of the indigenous populations in Florida and Georgia, with the coastal and St. Johns River chiefdoms particularly affected very early during the mission period. Both the coastal Guale and Timucua groups reached a 99 percent nadir early in the 17th century. Apalachee, while experiencing demographic issues of its own, was buffered to some extent from complete demographic collapse. The nadir of 73 percent is staggering but left a population base still capable of maintaining functioning communities, a leisure not afforded populations farther east.

Responses to impending demographic collapse were varied. The Spanish, ever promoters of their personal interests, saw colony maintenance as the most pressing issue. Widespread missions were indefensible and communication between them impossible, so critical links in the Spanish labor and travel network were maintained at all costs. The tiered process of population reorganization can be characterized as follows: (1) Stage 1 congregación involved the movement of local villages to the seat of the local chief, often where a doctrina had been established; (2) Stage 2 congregación involved the consolidation of previously distinct doctrinas, thus forming politically mixed mission communities; (3) Type 1 reducción was the forced movement of Christianized and semilocal populations to other locations within the mission chain; and (4) Type 2 reducción involved the in-migration of seminomadic, non-Christian, and nonlocal populations to deserted doctrinas, or for the purpose of establishing a new settlement with functional significance (to man a ferry, for example).

Predicted population genetic effects of such processes are difficult to verify and are based on the pattern of mate exchange that characterized precontact populations in the region. However, opportunities for mate exchange with nonlocal indigenous, African, and European populations would certainly have had an effect on levels of phenotypic variability in mission communities. If genetic drift was primary, a temporal decrease in genetic variability is expected. If aggregation and long-range gene flow were primary, a temporal increase in genetic variability would be expected. Patterning of increases, decreases, or stasis can be used to modify existing views on population definition and changes in population size. Of course, much of this discussion assumes a direct relationship between evolutionary and ethnohistoric processes and patterns of phenotypic variation. This relationship is further clarified in chapter 4.

Notes

1. In reference to this, McEwan records, "The Apalachees are the only known survivors of Florida's once numerous aboriginal populations. The Talimali Band of Apalachee Indians, who now reside in Louisiana, are still practicing Catholics, and their recent application for federal recognition is based largely on parish records" (McEwan 2001:642).

2. It should be noted that many Apalachee were scattered throughout the Southeast after the province was destroyed. The small number of Apalachee that remained Spanish loyalists was likely a small percentage of those who escaped into the woods, migrated west to French territory, or were captured by the English and returned to the Carolina colony as slaves.

3. As mortality rates increased throughout the contact period, it seems likely that the number of elderly individuals in the population decreased. Therefore, the ratio of effective to census population size would increase through time.

4. Worth (1992:155–158) records the presence of the village of Oconihad, which was a refugee settlement for individuals fleeing Timucua to avoid repartimiento labor draft requirements.

Part II
The Bioanthropology

4 Evolution and Transmission of Human Tooth Size

> Aside from general interest in the fascinating subject of early American archaeology, I found myself developing a deep special interest in a field which seems to have had little or no consideration from those engaged in such excavations. This omission is not at all surprising because archaeologists have not been informed of the tremendous importance of the skeletal and tooth and jaw specimens they have recovered, while the dental profession, for which such material opens an avenue of research concerning a vital present day problem, have been entirely unaware of the existence of such material.
>
> (Gillett 1927:291)

Because this book considers changes in phenotypic variation and relates such changes to evolutionary mechanisms and historical processes, the goal of this chapter is to establish the connection between patterns of odontometric variation and the social and demographic processes previously outlined. I address three topics in this chapter. First, the nature of phenotypic variation and its relationship to both genetic and environmental variation is discussed. Second, studies evaluating genetic inheritance of tooth size are summarized. Third, evolutionary mechanisms that affect levels of phenotypic variation in a population are outlined. This final topic adds greater insight into the mechanics of evolutionary change, given the historical details and processes discussed in chapter 3.

Nature of Phenotypic Variation

Phenotype (eye color, blood type, hair color, tooth size) refers to the physical expression of underlying genotypic variation (allele structure of the genes of an individual). Most models of biological distance assume that human osteodental phenotypic variation is subject to a polygenic, equal and additive effects model (Buikstra et al. 1990). That is, skeletal and dental morphology results from the input of multiple genes, each with multiple alleles that act additively to produce observed phenotypic form. Variation in allele structure among individuals produces variation in observed physical form, which is then "smoothed" by environmental, growth-related variation (Konigsberg 2000).

Figure 4.1 demonstrates how such a model is conceptualized. In this example, there are four different genes affecting the phenotype, each with three alleles.

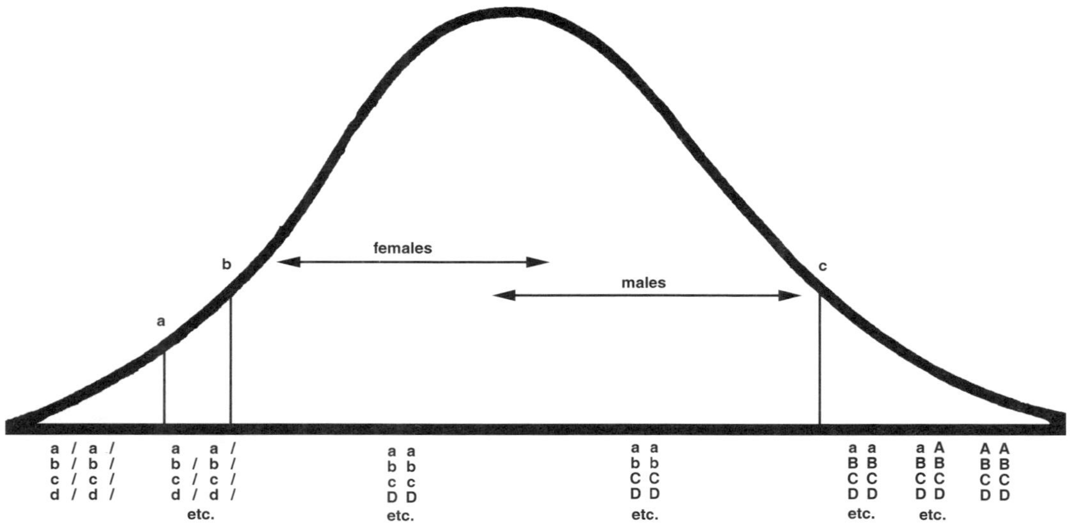

Figure 4.1. Equal and additive effects model for a phenotype (tooth size) controlled by four genes (A, B, C, D), each with three alleles (A, a, and a′, etc.). Double-headed arrows indicate the distribution of male and female phenotypes. Although there is some overlap, the degree of sexual dimorphism for tooth size is relatively high. An example of an allele structure for each is provided. Below the threshold value 'a' the tooth would have 0 form and be scored congenitally absent. Between 'a' and 'b' the tooth would form but would be much reduced in size, for example with a peg-shaped incisor or molar. The interval between 'b' and 'c' represents the "normal" range of variation in the population. Above 'c' the tooth would be abnormally large and may be considered hyperdontic. This model demonstrates how few genes with few alleles can act in an additive manner to increase the observed range of phenotypic variation.

Alleles represented by a lowercase letter produce a phenotypic response of one unit, uppercase alleles produce two units, and the lowercase prime (′) allele produces zero phenotypic units. Each individual has two alleles at each genetic locus (one paternal and one maternal) such that eight total alleles determine the genetic potential for tooth size. This model assumes there is no dominance of any form or epistasis.

If the assumptions are valid, this model easily explains dental phenotypic variation within a typical human population. Females tend to cluster on the left side of the distribution and males tend to cluster on the right side, leading to dental sexual dimorphism. In other words, females tend to have allele structures associated with fewer phenotypic units, that is, more lowercase and lowercase prime alleles. Below threshold 'a' the tooth would have form "zero" and be scored as congenitally absent (agenesis); between 'a' and 'b' the tooth would develop but be reduced in form and be scored as hypodontic (peg-shaped teeth, for example); between 'b' and 'c' the tooth would fall within the "normal" range of human variation. Above some arbitrary threshold 'c' the tooth might be scored as hyperdontic. Populations with smaller teeth have a greater proportion of lowercase and lowercase prime alleles, whereas populations with larger teeth have a greater proportion of uppercase alleles (or fewer lowercase prime alleles). Populations that are more variable have more represented alleles in appreciable frequencies. Populations that are less variable have fewer alleles or have all of the alleles but in disproportionate frequencies.

The simplicity of this presentation belies underlying complexity, however. Foremost is the limited number of possible phenotypes for a trait coded by four genes, each with three potential alleles. Because continuous phenotypes (i.e., measurements) can attain an infinite number of values, there must be another source of variation to explain this unlimited diversity.[1] While adding genes or more alleles for these genes to the model is one way to increase the potential range of phenotypic variation, environmental variation owing to growth disruption and other ontogenetic and environmental variables is also important (Konigsberg 2000). It is the relative importance of environmental or genetic variation that forms the core of most controversy surrounding paleogenetic approaches. Figure 4.2 demonstrates how the combined effects of genotypic and environmental variation work in concert to create infinite possible phenotypic outcomes and demonstrates the concept of "genetic potential," which may differ from observed tooth size.

The relative importance of additive genetic variance (that due to multiple genes acting additively) and environmental variance (that due to differences in childhood health, for example) for determining levels of phenotypic variability in a population is expressed as a numerical ratio (additive genetic variance/total phenotypic variance), termed heritability in the narrow sense (Hartl and Clark

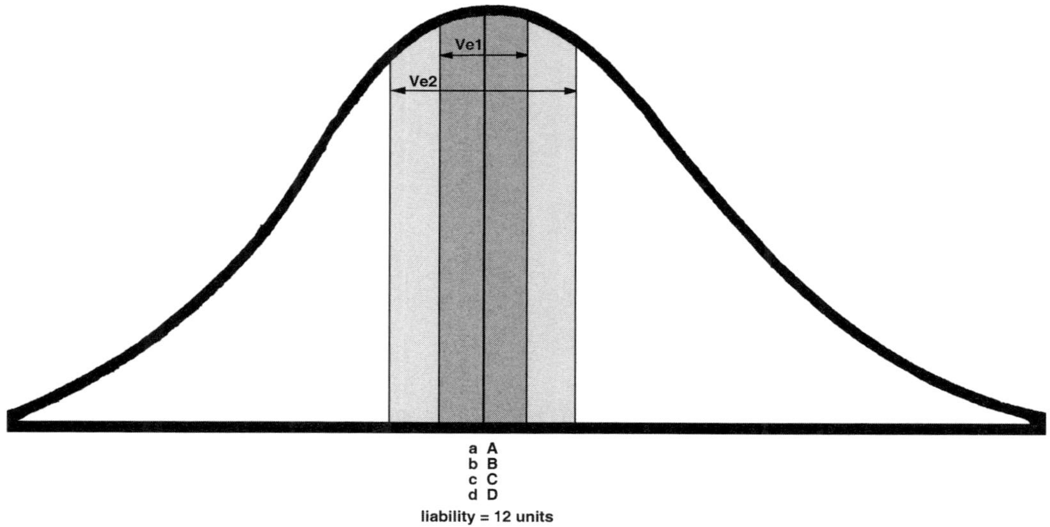

Figure 4.2. This figure depicts the underlying genetic liability for a single individual in a population. Whereas Figure 4.1 considered variation only in genotypes, Figure 4.2 adds the important environmental variance component. The individual has inherited a genetic liability of 12 phenotypic units for this trait. In one population, the environmental variance may be limited (Ve1) and has little effect on observed tooth form. In a second population, the environmental variance component could be large (Ve2), resulting in greater deviations from the inherited liability expectation of the trait. Because each individual will have a unique life experience, the addition of the environmental variance component greatly increases the observed range of phenotypes in a population.

1987). The "narrow sense" refers to the restriction placed on consideration of genotypic variance components; in this case only variation owing to additive effects (as per Figure 4.1) is considered because it is the only form of variance that can be transferred from parent to offspring. Dominance (the ability of an allele to mask the phenotypic effect of another allele) and epistatic (unlinked gene interaction) variance are ignored in this formulation. Narrow-sense heritability, hereafter simply heritability, varies between 0 and 1. A trait with 0 heritability indicates that none of the phenotypic variation for that trait *in that population* results from additive genetic variation. This is not the same as saying that the trait is completely determined by environmental variation. A trait with complete heritability (= 1) indicates that all phenotypic variation *in that population* results from differential underlying genetic liabilities for the trait. In this case, there is no environmental variation affecting phenotypic outcome. In reality, reported heritabilities for osteological data are often moderate in magnitude (Cheverud and Buikstra 1981a, b, 1982; Konigsberg and Ousley 1995; Relethford and Blangero 1990; Sjøvold 1984; Susanne 1977).

Misunderstanding and misinterpretation of heritability has resulted in overinterpretation of the significance of this statistic (Feldman and Lewontin 1975). Most crucial to this book is realization that 0 heritability does *not* indicate a lack of genetic control of the trait. Several points are worth noting (Falconer and Mackay 1996; Hartl and Clark 1987; Konigsberg 2000).

1. Heritability is not a measure of "how genetic" a particular trait is.
2. The heritability of a trait is not a condition of the trait itself but of the manifestation of variation of that trait within a population at a single point in time. Therefore, heritability is not invariant.
3. A phenotypic trait (such as tooth size) does not have a heritability but has heritabilities. More specifically, a particular trait has a different heritability within each population and for each generation within that population. Heritability is, therefore, a temporally and geographically static concept.
4. Causes of reported variation in heritability for the same trait are complex and relate to (a) levels of environmental variation in a population, (b) natural and artificial selection acting on a population, (c) allele frequencies within a population, (d) linkage disequilibrium for significant loci in a population, (e) dominance and epistatic (gene interaction) variation, and (f) population demography and age structure.
5. Heritability is not a measure of the resilience of a particular phenotype to environmental or ontogenetic growth perturbations (canalization). Traits with heritabilities close to 1 are not less prone to environmental modification.

On the other hand, heritability is useful for several things: (1) estimating predicted response to selection, for example in artificial breeding programs; (2) estimating predicted phenotypic correlations between relatives, despite tendencies of shared maternal and childhood environments to lead to overestimates of true heritability; and (3) assessing the selective neutrality of specific traits—traits with high heritability are typically unrelated to reproductive fitness. That is, selectively advantageous traits typically have 0 heritability because there is 0 genetic variation for this trait.

At this point it should be clear that interpreting heritability is a complex process. What does a heritability of 0 really mean? It can mean that (1) selection has been affecting the trait, (2) the population is entirely inbred (clones) or selection may have removed all of the additive genetic variation in that population, or (3) all phenotypic differences between individuals within the population are due to differences in environmental experience. It is important to note, however, that a low heritability does not mean that phenotypic form is unrelated to underlying genotypic values or that there is no genetic locus affecting the phenotype. Scott and Turner (1997) state this position most eloquently:

> This value [heritability] is not a measure of "degree of genetic determination." Say, for example, the development of trait X is controlled by genes at three loci, A, B, C. Assume further that all individuals in a population are homozygous at these three loci (i.e., AAbbCC). In such a population, any variation in trait expression is entirely environmental in origin—genetic variance and heritability both equal zero. An heritability of zero does not vitiate the fact that the development of the trait is controlled by genes—it is only the within group variation in trait expression that is determined by environmental factors. (Scott and Turner 1997:154)

A low heritability also does not mean that the trait is useless for differentiating populations in a phenetic sense (see Relethford 2004). This point is demonstrated by consideration of Figure 4.3 (adopted after Relethford 2004:Figure 1). This figure portrays genetic distances between three populations at two points in time. The length of the vertical bars indicates the amount of phenotypic variation in each population. The division of the horizontal bar by the vertical bar indicates the proportion of genetic (right side) and environmental (left side) contributions to overall variability. For time = 1, there is little environmental variance; most differences between populations are owing to additive genes, and the heritability of the trait is close to 1. Populations 'A' and 'B' are more closely related to each other than either is to 'C'. At time = 2, environmental variation has increased; heritability is now closer to 0 (within one generation), and more within-population phenotypic variability is owing to envi-

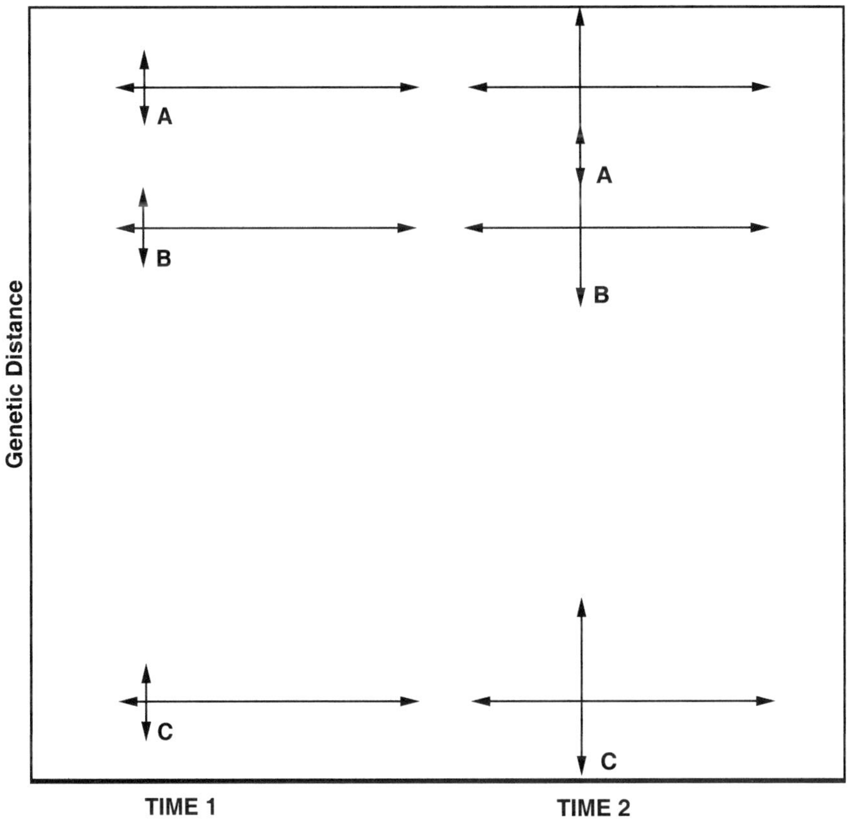

Figure 4.3. This figure demonstrates the relationship between total phenotypic variance, additive genetic variance, and environmental variance. At time = 1, three populations have similar levels of phenotypic variability as represented by the length of the vertical line. Most of the variation is owing to variation in additive genes (the right side of the horizontal arrow) and very little of the total variation is due to environmental variability (the left side of the horizontal line). Heritability is very close to 1. Populations A, B, and C are phenotypically distinct. At time = 2 the total phenotypic variability in each population has increased (vertical bar length increases) and the proportional contribution of environmental to genetic variation in each population has also increased. That is, environmental variation now contributes approximately 40 percent to overall phenotypic variability within each population. Note that A and B overlap slightly in the distribution of values owing to the increased variability but are still largely distinct. Both A and B are easily distinguishable from C, despite the dramatic reduction in heritability for this trait.

ronmental variation. However, genetic distances among populations 'A', 'B', and 'C' have not changed dramatically; 'A' and 'B' are still most similar, while 'C' is most divergent. The point at which environmental variation masks the between-group variation owing to genotypic values depends on two factors: how dissimilar the populations were at time = 1, and how severe the environmental variation is at time = 2. For example, while we have little trouble differentiating 'A' from 'B' from 'C' at time = 1, the environmental variation at time = 2 has confounded the phenotypic differences that existed between 'A' and 'B', making it more difficult to segregate individuals from these populations.

On the other hand, heritabilities close to 1 can also be interpreted in several ways: (1) the trait is not related to reproductive fitness; there is no selection for or against this trait; (2) the population is not inbred, so therefore there is sufficient genetic variability in the population; and (3) there is little environmental variability affecting phenotypes in the population. This does not mean all individuals in a population are phenotypically identical because narrow-sense heritability considers only the ratio of additive genetic variance to total phenotypic variance; dominance variance or epistasis could be operating to affect phenotypic form. And, in fact, traits demonstrating high heritability generally indicate that there is more, not less, overall variability in the population. In addition, large heritabilities do not guarantee that environment cannot affect the phenotype, only that it appears not to be significantly affecting phenotypic variation in the current population. In fact, heritability is primarily viewed as a measure of selection pressure:

> Heritability is related to the relative level of genetic variation, not to the strength of genetic control. . . . Selection progress eventually decreases the amount of genetic variation, and so, as selection progress is made, the heritability proceeds toward zero. However, the genetic control of the trait has not decreased, as shown by the improvement in the population that is the result of the genetic selection. This will emphasize that when a population has low or zero heritability for a trait, the correct inference is lack of genetic variation rather than lack of genetic control. (Mettler et al. 1988:176)

This final point is worth emphasizing. Heritabilities close to 1 speak well for this research design because it implies little selection, little environmental variation, and maximum additive genetic variation. Heritabilities close to 0 do not necessarily imply weakness because of the difficulty in interpreting the cause(s) of this low heritability estimate.With these caveats in mind, the discussion turns to a consideration of known mechanisms of dental size inheritance patterns and previous estimates of narrow-sense heritability.

"What Big Teeth You Had, Grandma"[2]

Most published research suggests that dental size variation conforms to a polygenic, equal and additive effects model of inheritance with limited evidence for dominance or pleiotropy (Bailit 1975; Bowden and Goose 1969; Dempsey et al. 1995; El-Nofely and Tawfik 1995; Kolakowski and Bailit 1981; Potter et al.

1976; Townsend and Brown 1978a, b). In addition, determining whether genes for crown size are autosomal has received considerable attention in the literature, using two different approaches: estimating size correlations between siblings and parents and comparing these to theoretical values based on models of X or Y linkage, and comparing tooth sizes in individuals with chromosomal abnormalities to nonaffected comparative samples. The former approach is based on theoretical expectations of shared chromosomes. For example, in an X-linked trait, the father-son correlation is expected to be lower than all other combinations because they never share a common X chromosome. For sibling comparisons, the correlation between sisters should be greater than between brothers which, likewise, should be greater than brother-sister correlations. Garn et al. (1965a, 1967a; Garn, Lewis, and Wallenga 1968) found sister-sister correlations exceeded brother-brother correlations which exceeded brother-sister correlations, supporting an X-linked model of inheritance for crown shape and crown size. Lewis and Grainger (1967) and Alvesalo (1971) produced similar results. Sexual dimorphism in tooth size suggests Y chromosome involvement as well. Bowden and Goose (1969), Hu et al. (1991), Niswander and Chung (1965), Potter et al. (1968), Potter et al. (1983), and Townsend and Brown (1978b), however, found no evidence for higher sibling correlations by sex, suggesting a predominantly autosomal mode of inheritance.

Chromosomal abnormality data similarly suggest a nonautosomal contribution to tooth size. Alvesalo and coworkers found that 47 XYY males (Alvesalo and Kari 1977; Alvesalo et al. 1975; Townsend and Alvesalo 1985a) and 47 XXY males (Alvesalo and Portin 1980; Alvesalo et al. 1991; Townsend and Alvesalo 1985b) exhibit larger crown dimensions than 46 XY males. Likewise among females, 45 X (Filipsson et al. 1965; Kari et al. 1980; Mayhall and Alvesalo 1992; Townsend et al. 1984), 45 X/46 XX (Varrela et al. 1988), and 46 Xi (Mayhall et al. 1991) females exhibit smaller crown dimensions than 46 XX females. Females with testicular feminization syndrome (46 XY) have dimensions comparable to 46 XY males (Alvesalo and Varrela 1980; Grön and Alvesalo 1995).

Heritability of Tooth Size

Studies of dental inheritance have a long history owing primarily to clinical applications for treatment of malocclusion and orthodontic problems. Earlier research focused on twin and triplet comparisons as a predictor of orthodontic malocclusion. Although no heritability estimates were presented, the general finding was that identical twins had similar arch profiles, eruption timing, and crown dimensions (Bachrach and Young 1927; Braun 1938; Korkhaus 1930; Newton 1937; Potter and Nance 1976; but see Cohen et al. 1942; Wood and Green 1969). Concordance was so high that Lundström (1948) declared tooth profiles a suitable means for testing twin monozygosity. Initial interpretation of these results was that arch size, eruption timing, and dental dimensions were under strong genetic control. However, because twins share similar environments for much of their lives (beginning *in utero*), there is uncertainty about the relative contribution of genes to phenotypic form. According to Potter et al.,

> in quantitative genetics, the most careful evaluation of twin studies may yield only the broad conclusion that persons who are identical in their genotype are more likely to react similarly to their environments than are persons with dissimilar genotypes. Evidence for significant genetic control of variability in tooth size from the twin method unfortunately does not satisfactorily delineate the inheritance of tooth size for predictive purposes in clinical orthodontics or microevolutionary analysis. (Potter et al. 1968:89)

Christian and coworkers (Christian 1979; Christian and Norton 1977; Christian et al. 1974) developed statistical techniques to correct for problematic assumptions of previous twin studies (differences associated with mono- or dizygosity are purely genetic, environmental covariation is equal regardless of zygosity, and assumed association between zygosity and trait mean values—Sharma et al. 1985). Later studies using better methodologies produced more variable results:

(1) Corruccini and Potter (1980) generated an overall arch size heritability of 0.27.
(2) Townsend et al. (1986) estimated mesiodistal heritabilities for upper incisors with an average of 0.30 and a maximum of 0.42.
(3) Sharma et al. (1985) estimated mesiodistal and buccolingual heritabilities for the dentition, averaging 0.60.
(4) Dempsey et al. (1995) generated upper and lower incisor heritabilities averaging 0.86 with a maximum of 0.91.
(5) Harzer (1995) estimated tooth size heritabilities of 0.53–0.78 for the maxillary teeth and 0.46–0.55 for the mandibular teeth.
(6) Townsend et al. (2003) estimated crown size heritabilities ranging between 0.60 and 0.82.
(7) Dempsey and Townsend (2001) generated crown size heritabilities ranging from 0.56 to 0.92, with most estimates greater than 0.80.

Heritability studies based on extended pedigrees eventually came to replace those based on twin data. Both Dockrell (1956) and Garn et al. (1965a) used correlations between relatives as a measure of genetic inheritance; estimates were around 0.80. Other studies were to follow:

(1) Bowden and Goose (1969) estimated maxillary anterior tooth heritabilities, averaging 0.50.
(2) Alvesalo and Tigerstedt (1974) estimated mesiodistal and buccolingual heritabilities for 40 dental dimensions based on 90 groups of Finnish siblings. Thirty-two of the heritabilities were greater than 0.50 with an overall heritability of 0.59. The average mesiodistal heritability was 0.59 and the average buccolingual was 0.67. The maxillary dimensions averaged 0.67 and the mandibular .51.
(3) Alvesalo and Tigerstedt (1974) calculated heritabilities from Lundström's (1948) data set. The average estimate was 0.78; maxillary metric heritabilities averaged 0.80 and the mandibular .77.
(4) Townsend and Brown (1978a) estimated full-sib crown size heritabilities, averaging 0.72 for mesiodistal dimensions and 0.81 for buccolingual dimensions. Half-sib results were much lower, 0.63 for mesiodistal diameters and 0.31 for buccolingual diameters. They estimated that 52 percent of the variance in dental size was additive in nature, 12 percent was owing to common environment, and 36 percent was owing to within-family effects.
(5) Harris and Smith (1980) estimated heritabilities averaging 0.60 for multiple crown dimensions.
(6) Kolakowski and Bailit (1981) generated heritabilities of 0.23 for buccolingual anterior tooth dimensions, 0.96 for mesiodistal anterior tooth dimensions, 0.66 for premolar size, and 0.38 for molar size. Complex segregation analysis that should remove some of the environmental covariation returned estimates of 0.001, 0.806, 0.614, and 0.558, respectively.
(7) Harris and Smith (1980) estimated arch width heritabilities varying from 0.48 to 0.77.
(8) Harris and Smith (1982) found arch size to have a narrow-sense heritability of 0.56.
(9) Potter et al. (1983) found that 52 percent of the first molar variance and 35 percent of lateral incisor variance were owing to transmissible factors (though not necessarily all of them were genetic).
(10) Hu et al. (1991) reported heritability estimates of 0.93 for arch perimeter, 0.83 for arch width, 0.68 for arch length, and 0.74 for a general size index. They attributed about 80 percent of the total variation in their data to additive genetic components and found that anterior tooth dimensions, particular in the mandible, were more subject to environmental effects than the posterior dentition.

Despite careful analytic techniques designed to parcel the effects of genetic and environmental variation, the nature of environmental covariation is still uncertain. According to Potter et al., "Dental size is a trait for which the level of familial resemblance is high, but for which biological and environmental correlates are not yet understood" (Potter et al. 1983:284).

Both twin and family studies suffer the shortcoming that shared maternal and childhood environments result in similar phenotypes for nongenetic reasons. As noted by King et al. (1993), siblings share both a common intrauterine environment and postnatal lifestyle and therefore postnatal factors such as diet, nutrition, socioeconomic status, and disease loads all push siblings along a similar epigenetic pathway. According to Potter and colleagues, "The generally higher correlations for sib-sib than for parent-offspring . . . further reflect strong environmental effects on tooth size factors, since sibs are more likely to share a common environment than are parent and offspring" (Potter et al. 1968:97).

The preponderance of data from studies of dental genetics supports the assumption that tooth size is polygenic and conforms to an equal and additive effects model of inheritance and expression. Heritabilities vary widely depending on the population studied and the methodology used to generate the estimates; however, taking a rough average of all the cited research (Alvesalo and Tigerstedt 1974; Corruccini and Potter 1980; Dempsey et al. 1995; Goose 1971; Harris and Smith 1980; Potter et al. 1983; Rebich and Markovic 1976; Townsend and Brown 1978a, b) suggests a mean heritability of approximately .62, roughly 7 points higher than similar estimates for craniometric and anthropometric data (Devor et al. 1985; Konigsberg and Ousley 1995; Relethford and Blangero 1990).

With full understanding of what heritability means and establishing that tooth size does typically generate moderate to large heritability values, we are now in the position to link the evolution of dental phenotypes to specific mechanisms that complement the ethnohistorical processes outlined in chapter 3.

The Four (Minus Two) Mechanisms of Evolution

Evolution results from a change in allele frequencies through time, which, under an equal and additive effects model, leads to concomitant changes in phenotypic means and variances within a population. Classic expositions recognize four mechanisms (not forces in the sense of gravity) that effect allele frequency change: mutation, natural selection, genetic drift, and gene flow (Table 4.1). Mutation is not considered in this work because mutation rates are very low and are ineffective mechanisms of evolutionary change over short periods of time. In addition, I remain absolutely unconvinced that tooth size was in any way related to reproductive fitness over five generations during the 16th and 17th centuries. As discussed, tooth size generally demonstrates moderate heritability values, which suggests that it is unrelated to fitness and is therefore

Table 4.1. Summary of evolutionary mechanisms

Mechanism	W/I pop variance change	B/T pop variance change	Basis	Scale
Mutation	increase	increase	new alleles created	Macro
Natural selection	both	increase	environmental adaptation	Macro
Genetic drift	decrease	increase	stochastic sampling	Micro
Gene flow	increase	decrease	mate exchange	Micro

Note: W/I = within, B/T = between

selectively neutral (over short time periods). Also, there is little reason to believe people were dying during the mission period because of deficiencies (or excesses) in tooth size, particularly given that epidemics and warfare were largely responsible for mortality and the unlikelihood that tooth size is genetically linked with markers of immune response and disease susceptibility. One could envision selection for larger body size in the mission environment, particularly among males for whom work load requirements were particularly brutal, and such models have been proposed to explain muscularity and body mass in Oceanic populations (see, for example Houghton 1996). However, tooth size in humans (but not other primates) is poorly correlated with body size (Filipsson and Goldson 1963; Henderson and Corruccini 1976), making the connection between them problematic. Therefore, despite the position of natural selection and mutation as the most integral long-term evolutionary mechanisms affecting species diversity, the brief temporal scope of the sampling design and the tenuous correlation between tooth size and fitness suggest that other mechanisms take precedence. These mechanisms are genetic drift and gene flow, the primary determinants of population structure in humans (Relethford 1980).

Genetic Drift

Genetic drift reflects the effects of stochastic and imperfect sampling of alleles such that minor variation in allele frequencies define subsequent generations. Genetic drift leads to a reduction in genetic variability within populations but increases genetic variability between subdivided populations. The decrease in within-population variability results from allele fixation or loss, depending on the initial allele frequency in the population. For example, an allele at 10 percent frequency is more likely to become lost in the population (0 percent), whereas one at 90 percent is more likely to become fixed in the population (100 percent). Once an allele has become fixed, there is no variation for that gene in the population, natural selection is completely ineffective at evolutionary change, and the only way to increase genetic variability is via mutation (a new allele is created) or gene flow (a new allele is introduced into the population through interbreeding). At the same time that drift leads to a decrease in within-group variability, it also increases variability between populations. This reflects the random nature of drift. Although there is a tendency for high-frequency alleles to become fixed and low-frequency alleles to become lost, this does not have to be the case. And, for loci with alleles in appreciable and moderate frequencies (40–60 percent), the effects of drift can be very unpredictable and random. The net effect is that gene frequencies tend to change in unpredictable and often different directions in subdivided populations.

The final feature of genetic drift with importance to this study is the relationship between drift effects and population size. Drift is of limited consequence in very large populations (such as those of today) and is a most effective mechanism of evolutionary change in small populations. To demonstrate why, I consider a simple coin-flipping experiment. If you flip a coin ten times, sometimes you will get six heads and four tails, four heads and six tails, or even the rare occasion when nine heads and one tail are produced. These rare events (9:1, 10:0 ratios) are more likely to occur as the numbers of trials or experiments (or in our case subdivided human populations) increase. Even the relatively innocuous case of a 6:4 result causes a significant deviation (10 percent) from the expected ratio (5:5) because the number of trials was so small. If we extend this example to 1,000 flips, then the ratios are going to be much less aberrant, with most trials resulting in a ratio in the range of 450:550, and vice versa, or a 5 percent random fluctuation from expectation (500:500) on average. A 999:1 ratio of heads to tails is very unlikely. In human terms, in small populations there will be some alleles that are not represented in the next generation merely by chance alone. And, the net effect of this chance sampling process is much greater in smaller populations. In small populations, the time it takes for an allele to become fixed or lost is minimized. In large populations, the sampling error results in much smaller absolute deviations from expectation, and the probability of a dramatically aberrant event is very small.

Gene Flow/Migration

Gene flow is the movement of genes or people across the landscape leading to interaction and interbreeding between different populations. Gene flow increases genetic variability within a population in the short term and decreases variation between populations in the long term. That is, the introduction of a new allele into a population will increase genetic variability in that population. However, over long periods of time, gene frequencies in two populations that were previously distinct (owing primarily to drift or natural selection) reach equilibrium, and we may consider the populations to be undifferentiated; two different populations no longer exist. Therefore, the primary role of gene flow is to redistribute genetic variation that exists within a species.

As alluded to in chapter 3, the importance of changes in population size (drift) is mitigated by the changing definition of that population (gene flow). In most species, including humans (but this is becoming less so the case), geography is the primary determinant of patterns of gene flow. Individuals from populations physically distant from one another, such as Guale and Apalachee, are less likely to mate. Less problematic for humans are physical barriers to mate exchange over short distances. Human technology mitigates the effects of mountains, rivers, and lakes that figure so prominently in basic expositions on reproductive isolation. Humans, however, suffer the additional consideration of cultural and social prescription and proscription. That is, cultures define normative rules of mating behavior that are codified within a kinship system, which are, nonetheless, often ignored (Moore 1994a, b). At the same time, cultures inhibit mate choice via various markers of identity, such as class, caste, kin group, language, political structure, religion, and economic class, among others. Therefore, investigation of patterns of gene flow, while biological in implementation, is wholly cultural and social in interpretation.

A Synthetic Model of Bioethnohistory

We can now combine discussions in chapters 3 and 4 into a unified model of evolutionary change with full consideration of the ethnohistoric and evolutionary genetic data. For this, I reconsider the 4-loci, 3-allele model discussed above.

Polygenic Inheritance and Tooth Size Variation

Phenotypic variability increases when the number of genotypic combinations of alleles increases. Table 4.2 presents five different scenarios using the 4-gene, 3-allele system. Populations 1–3 consist exclusively of homozygotes, and all of these populations are completely invariant, that is, everyone within the population has exactly the same tooth size. Population 1 consists of large-toothed individuals, population 2 consists of small-toothed individuals, and population 3 consists of agenetic (edentulous) individuals. Note that all three populations are distinct phenotypically and exhibit very different mean dimensions (16, 8, 0 units), but they have identical variance measures (0). Note that heritability is 0 in all three populations, but phenotypes *are* determined completely by genes. If population 1 actually varied between 15.9 and 16.1, population 2 between 7.9 and 8.1, and population 3 between 0 and 0.1, then all variation within a population would be environmental in origin and heritability would still be 0. To reiterate, 0 heritability does not mean there is no genetic contribution to a phenotype.

Table 4.2. Polygenic model of equal and additive effects

Population	Alleles	Phenotypes	Variability
1	A, B, C, D	all big teeth (16 units)	Minimized - 0 heritability
2	a, b, c, d	all small teeth (8 units)	Minimized - 0 heritability
3	a′, b′, c′, d′	edentulous (0 units)	Minimized - 0 heritability
4	A, a, B, b′ C, c, c′, D	numerous, small, large, moderate teeth	Moderate
5	A, a, a′ B, b, b′ C, c, c′ D, d, d′	numerous, many combinations of small, moderate, and large teeth	Maximized

Population 5 represents the other end of the spectrum. All three alleles for each locus are present within this population (12 total alleles), and multiple tooth sizes that span the range of potential variation (from 0 to 16 units) are observed. This population represents maximum variability under this inheritance system owing to the number of genotypic combinations. Population 4 is polymorphic for each loci but lacks some of the alleles represented in population 5. Phenotypic variability is diminished as a result and would be intermediate in magnitude.

The discussion above demonstrates that more alleles equal greater phenotypic variability. Populations lacking variation in allele presence are variance-restricted as a result. However, the frequency of individual alleles is also an important determinant of tooth size variability because the frequency of heterozygotes is maximized when allele frequencies approach 50 percent (for a 2-allele locus).[3]

In Table 4.3, I consider two populations in which all 12 alleles are present (in other words, similar to population 5 in Table 4.2). In population 1, allele frequencies are divergent with a predominance of the capital allele for each locus. Because of the disparity in frequencies, and the relative rarity of the lowercase and lowercase prime alleles, most individuals in this population will be homozygous for the uppercase alleles. This indicates that most individuals will have large but similar tooth size. Variability is, therefore, not high. In population 2, all alleles are represented in appreciable frequencies, resulting in equal numbers of multiple genotypes. This population has smaller average tooth size than population 1 but exhibits greater variability. Therefore, both the alleles present and the frequencies of each allele are important determinants of phenotypic mean and variance statistics.

Table 4.3. Relationship between genotypic and phenotypic variability

Alleles present	Population 1 frequency	Population 2 frequency
Gene 1		
A	98	33
a	1	33
a′	1	33
Gene 2		
B	98	33
b	1	33
b′	1	33
Gene 3		
C	98	33
c	1	33
c′	1	33
Gene 4		
D	98	33
d	1	33
d′	1	33

Genetic Drift and Phenotypic Variability

It was noted above that small populations experience the effects of genetic drift in more dramatic fashion. To model this, I consider gene 1 from population 1 listed in Table 4.3. If mating is random and the sex ratio is equal, then the probability of choosing any single allele is equal to the frequency of that allele. Assuming a small population size of 50 individuals (with 100 total alleles), the probability of selecting an 'A' allele is 98 percent and an 'a' or 'a′' allele is 1 percent, respectively. The probability of selecting two 'a' or 'a′' alleles (one from a father and one from a mother) is 4 percent, an unlikely event. Conversely, the probability of *not* picking either an 'a' or 'a′' allele is equal to the probability of selecting two 'A' alleles, which is 96 percent. Therefore, for each generation there is a 96 percent chance of the 'a' and 'a′' becoming lost, resulting in the complete lack of genetic variability for this gene. If only one person survives to the next generation (a population bottleneck caused by an epidemic, for example), then there is a high probability that the 'a' and 'a′' alleles will not be represented in this person. If population size does not decline, then the probability of faithful representation of all alleles increases. If the preceding were repeated for each of four loci, then, all else being equal, all populations would eventually resemble populations 1–3 in Table 4.2. That is, genetic drift works to reduce genetic variability and increase homozygosity within a population.

Changes in population size lead to similar changes in the expected effects of genetic drift. And, because the probability of allele fixation is equal to the initial allele frequency (Hartl and Clark 1987), the smaller the population size, the greater the potential effect of any given allele. For example, if there is only one copy of the 'a' allele in a population of size 50, there is a 1 percent chance of that allele becoming fixed, an unlikely occurrence. However, if the size of this population decreased to 5 individuals (and the 'a' allele happened to be carried by one of these individuals—the randomness factor), then the potential effect of this rare allele is much greater. The probability of allele fixation has increased to 10 percent. Also recognize that if we considered 100 different populations, each with initial allele frequencies of 'A' equals 98 percent and 'a' and 'a′' equal 1 percent, respectively, then we expect one population to become fixed for 'a', another to become fixed for 'a′', and 98 to become fixed for 'A'. In all 100 populations, drift is working to fix one of these alleles, thus reducing heterozygosity. This explains why subdivided populations (Apalachee and Guale) are expected to diverge (have different allele frequencies) under a pure drift model. Genetic drift can be summarized as follows:

(1) Genetic drift reduces variation within a population. The underlying mechanism for variance reduction is the fixation or loss of alleles at a locus.
(2) Genetic drift increases variation between subdivided populations. The underlying mechanism is the stochastic nature of the allele sampling process, thus leading to expected divergent allele frequencies in subdivided populations.
(3) Smaller populations experience greater drift effects because heterozygosity is lost more quickly as alleles deviate and approach fixation or loss.
(4) Decreasing population size results in a loss of genetic variability as a result.

Gene Flow and Phenotypic Variability

To model gene flow, I once again consider the two populations in Table 4.3. In Table 4.4, I consider the effects of gene flow on allele frequencies resulting from individuals from population 2 (e.g., Yamassee, Guale, Timucua) migrating into population 1 (e.g., Apalachee). In the first generation, the allele frequencies for the 'A', 'a' and 'a′' alleles are 98, 1, and 1 percent, respectively (from Table 4.3, population 1).

In the second generation, ten randomly selected migrants from population 2 migrate into population 1, resulting in the change of alleles frequencies presented for generation 2. In the second generation the allele frequencies for the 'A', 'a' and 'a′' alleles are 87, 6.5, and 6.5 percent, respectively. The rate of change

Table 4.4. Modeling effects of gene flow on allele frequencies

	Generation 1			Generation 2		Generation 3		
	Allele		Migrants	Allele		Migrants	Allele	
Alleles	Freq.	No.	from P2	Freq.	No.	from P2	Freq.	No.
A	98%	49	+3.3	87%	52.3	+6.6	74%	58.9
a	1%	.5	+3.3	6.5%	3.8	+6.6	13%	10.4
a′	1%	.5	+3.3	6.5%	3.8	+6.6	13%	10.4

for each allele is −11 percent, +5.5 percent, and +5.5 percent, respectively. Variation has increased within population 1 (Apalachee) because the proportion of heterozygous genotypes has increased, and the variation between populations 1 and 2 has decreased because allele frequencies are more similar in this generation than in the previous.

In the third generation, 20 randomly selected individuals from population 2 migrate into population 1. The allele frequencies for the 'A,' 'a,' and 'a′' alleles are 74, 13, and 13 percent, respectively. The rate of change for each allele is −13 percent, +6.5 percent, and +6.5 percent, respectively, which is greater than that demonstrated between generations 1 and 2. Increasing the number of migrants, therefore, increases the rate at which allele frequencies between two populations reach equilibrium. Continuation of this process of migration would eventually lead to a convergence of allele frequencies in these populations with the convergence rate determined by the rate of migration and the degree of initial difference. Conflation of two minimally divergent Native American populations would result in rapid convergence of allele frequencies. This may not be the case for Spanish or African migrants into the Apalachee and Guale populations, however.

Effects of Spanish or African admixture on Native Apalachee or Guale phenotypic diversity can be modeled simplistically by considering populations 1 and 2 from Table 4.2. Assuming Native Americans have larger teeth than Spaniards (see Dittmar et al. 1998; Kieser 1990), we can assign Native Americans to population 1 (all large teeth) and Spaniards to population 2 (all small teeth). Combining the two populations and assuming random mating between males and females of both populations would quickly lead to a dramatic increase in phenotypic variability. Whereas in isolation there was only one phenotype in each population, admixture between the two results in multiple intermediate tooth size phenotypes. Recognition of the initial distinctiveness of the populations is paramount because if two populations that were self-identified as different groups (for example, Apalachee and Yustaga) had similar alleles and frequencies of those alleles, then the result of gene flow between them would lead to no increase in within-population phenotypic variability. This fact motivated much of the discussion in chapter 3. The primary effects of gene flow can be summarized as follows. (1) Gene flow increases within-population variation and decreases between-population variation over the short term. (2) The rate of increase in within-population variation and decrease in between-population variation is a function of the migration rate and amount of gene flow between two populations. (3) The mechanism of change within a population is either the introduction of a new allele from an external source or the increase in frequency of rare alleles, leading to a greater probability of heterozygosity.

Parting Words of Caution

Human paleogenetic research has experienced, and continues to experience, its share of controversy, ranging from accusations of adopting a racial and typological perspective (Armelagos and Van Gerven 2003) to arguments over the role of environment in determining phenotypic form (Houghton 1996). The latter has inspired recent reappraisals of Boas's famous immigrant study (Gravlee et al. 2003; Relethford 2004; Sparks and Jantz 2002). Accusations of racialism (the use of race as a tool, which differs from the implied hierarchy of racist perspectives) are not of concern to this study because focus on variation is nontypological by definition. However, consideration of the potential effects of environmental variation (as discussed above) is a concern. My position on this issue can be summarized as follows: (1) odontometric data have proven more heritable than other forms of paleogenetic data; therefore, odontometric data minimize the potential effects of environment on phenotypic variance estimates; (2) assuming that environmental variance was significantly affecting phenotypic form in the *La Florida* populations does not, by default, nullify interpretations of a genetic nature. Because sampling is within a lineage at three points in time, proposing environmental variation as the source for differential expressions of phenotypic variance also assumes interindividual heterogeneity in the mission experience. While it is easy to argue that metabolic stress increased during the contact period (Larsen 2001), it is more difficult to assume that some indigenous individuals *within a population* were selectively protected from such stress. It seems most likely that such events would have affected measurement values (means) but not necessarily phenotypic variances. That stress would diminish genetic potential due to metabolic disturbance during the period of dental development contradicts previous research documenting a secular increase in tooth size in the mission populations (Stojanowski 2004). This suggests the phenotypic effects of childhood stress are not so straightforward.

Investigating patterns of tooth size variation in relationship to social and evolutionary attributes during the Spanish colonial period is preferential for several reasons: (1) because enamel is the hardest substance in the human body, it preserves much better than osseous tissues; (2) studies of genetic transmission and heritability indicate a comparatively large additive genetic variance component to phenotypic expression in multiple population groups, which relates to (3) amelogenesis is not a process of tissue maintenance but of tissue formation; once the crown has formed it does not undergo continuous remodeling during the course of one's life (as the craniofacial complex does), and therefore the length of time for environmental variation to manifest and affect phenotypic variation is minimized; and (4) because teeth form from the tip of the crown to the root, crown morphology is established very early in an individual's life, further reducing the exposure of odontometric data to environmental variation.

Four classic mechanisms affect levels of genetic/phenotypic variability within a population. Mutation is the physical change in the structure of a gene, which is ultimately responsible for the formation of new alleles and thus increases genetic variability within a population. Natural selection affects allele frequencies through the preferential and differential selection of individuals, owing to inherited traits that confer an advantage to individuals expressing the trait. Neither mutation nor natural selection is likely to be a significant explanatory mechanism, given the short duration of the study period. To the contrary, genetic drift (a stochastic sampling process which decreases genetic variation owing to allele fixation and loss) and gene flow (mate exchange that increases within-population genetic variation and decreases between-population genetic variation) are both mechanisms that likely affected levels of phenotypic variability in mission populations. While genetic drift is related specifically to demographic variables (population size), patterns of gene flow are responsible for defining the boundaries of the population and reflect cultural and social norms and proscriptions.

Having fully developed the research design and related historical processes with evolutionary mechanisms, I am now in a position to present the research methodology. Specifics of data collection, pre-analysis data treatments, and statistical methods are discussed in chapter 5.

Notes

1. Theoretically, any continuous variable can attain an infinite number of measurements regardless of the limits applied to the range.

2. After Brace (1991).

3. Under the Hardy-Weinberg model, the frequency of heterozygotes is equal to 2pq, where p is the allele frequency of one allele and q is the frequency of the second allele in a two-allele system. This quantity is maximized when $p = q = 0.5$. Because heterozygosity represents variability, a population will exhibit maximum phenotypic variability when the frequency of heterozygotes is also maximized.

5 Conceptual and Research Methods

Chapter 5 presents the research methodology and outlines strategies adopted to reduce error and increase the resolution and veracity of the analytical results. Methods for dental phenotypic data collection are discussed as well as justification for using the particular battery of metric variables. Pre-analysis data treatments designed to minimize the effects of random and environmental error are then presented. Results of these analyses immediately follow. The chapter closes with discussion of the statistical methodologies adopted to evaluate the working research hypotheses. Mathematical formulae are presented in the appendix.

"30 White Horses on a Red Hill"

For reasons unknown to me, dental metrics seem to receive short shrift in the bioarchaeological literature. They do, however, perform equally as well as other classes of data in evolutionary research programs (Falk and Corruccini 1982; Kieser 1990) and offer several advantages over cranial data (both metric and nonmetric) and dental morphology.

Preservation

When I began data collection, there was little choice in the matter of which type of data to collect. And, if there was some doubt, this was quickly dispelled when I visited the bioarchaeology lab at the University of North Carolina–Chapel Hill where a majority of the research collections were on loan to Clark Spencer Larsen. As would most researchers, I decided to analyze the largest sample first, Santa Catalina de Guale, which represented just under 400 individuals. When presented with four small boxes filled with thousands of teeth and enamel caps, I knew odontometry was the only way to address the research questions. Therefore, the choice of dental metrics as a phenotypic proxy for population genetic variation was one of both necessity and choice.

Because dental enamel is the hardest substance in the human body, it preserves at a much higher rate than other morphometric variables of the craniofacial skeleton. Both the coastal (sandy) and inland (acidic clay) geological contexts in which the missions were established result in marginal skeletal preservation, even for samples only a few hundreds years old. The preservation of some of the mission material (e.g., San Luis de Talimali and Santa Catalina de Guale) was so poor that only enamel crowns survived; most dentine and root material had been destroyed, a pattern I also observed during church excavations at the O'Connell Mission site in Leon County, Florida (see Marrinan et al. 2000). Although some precontact samples exhibited moderate degrees of cranial preservation, a majority of the precontact comparative samples and all but one of the mission period samples were represented primarily by loose teeth and cranial fragments. Even where cranial preservation was good, intentional deformation and soil distortion created problems for traditional craniometric approaches. Therefore, the dentition was the only means for achieving moderately large and representative sample sizes. I would like to stress, however, that use of the dentition was not a decision based on necessity alone. As detailed below, odontometrics offer several distinct advantages over other variable types.

Ease of Measurement

Dental metrics (and metric data in general) are easier to record accurately than nonmetric or morphological variation of discontinuous scale (Kieser 1990). These data are, therefore, more easily replicated. With little practice or instruction, an inexperienced observer can record metric data much more accurately than is possible for dental morphological variation, which is subject to considerable inter- and intra-observer error.[1] Differences in laboratory lighting, magnification requirements, and microclimatic effects (such as humidity) are of more limited concern for the odontometrist.

Statistical Tractability

In addition to ease of observation, continuous data are also much easier than discontinuous data to manipulate and analyze using basic statistical applications. These applications include pre-analysis treatments and adjustments as well as univariate and multivariate inferential statistics and multivariate modeling. Statistical tractability results, in part, from the better documented underlying genetic structure of continuous variation. Continuous phenotypic variation manifests from a simple polygenic mode of inheritance, whereas discontinuous phenotypic variation conforms to a more abstract "threshold" model (Falconer 1981; Hartl and Clark 1987; Hauser and De Stefano 1989; Scott and Turner 1997). Because the underlying genetic model is better understood, metric data

Table 5.1. Age of formation for the mesiodistal and buccolingual dimensions of the polar teeth

Tooth	Crown Begins	Crown Completion	Mesiodistal formed Location[a]	Mesiodistal formed Age[b]	Buccolingual formed Location[a]	Buccolingual formed Age[b]
UI1	3–4 months	3.5 years	90%	1	10%	3.5
UC	4–5 months	4.5 years	70%	3	10%	5
UP1	1.5 years	5.5 years	75%	4.5	60%	6
UM1	at birth	2.5 years	50%	1.5	50%	1.5
LI2	3–4 months	3.7 years	90%	1	10%	4
LC	4–5 months	4.5 years	60%	2	10%	4
LP1	2 years	5.5 years	75%	5	60%	6
LM1	at birth	2.5 years	50%	1.5	50%	1.5

[a]This is determined as an approximate percentage of the total crown height as measured from the cemento-enamel junction. For example, a measurement typically located at 25% of crown height is closer to the cemento-enamel junction than one measured at 75% of crown height.

[b]Age in years at which crown reaches this point of formation. Data after Ubelaker 1978.

are more easily incorporated into modern population genetic analytical frameworks (Relethford 2003; Steadman 2001).

Ontogenetic Plasticity

Amelogenesis (the process of enamel production) ceases after initial crown formation (Hillson 1996). Once formed, dental crowns cannot repair or remodel during the course of one's life. While problematic in today's carbohydrate-rich, cariogenic environment, aplastic morphology significantly reduces the environmental variance component of dental phenotypic variation, quite unlike craniofacial metric and nonmetric variation, which is age-transgressive until death. This benefit is compounded by considering the age at which crown dimensions typically recorded in bioarchaeological analyses are formed (Table 5.1). The earliest crown completion occurs for the first molars with both mesiodistal and buccolingual (see Figure 5.1) dimensions completed by the second year of life. The buccolingual dimensions of canines and premolars are the last to form, with completion by the sixth year of life. Adding to these estimates the typical gestation length and the period for which odontometric variation is subject to growth perturbations, ontogenetic variation can be estimated at between three and seven years in total.

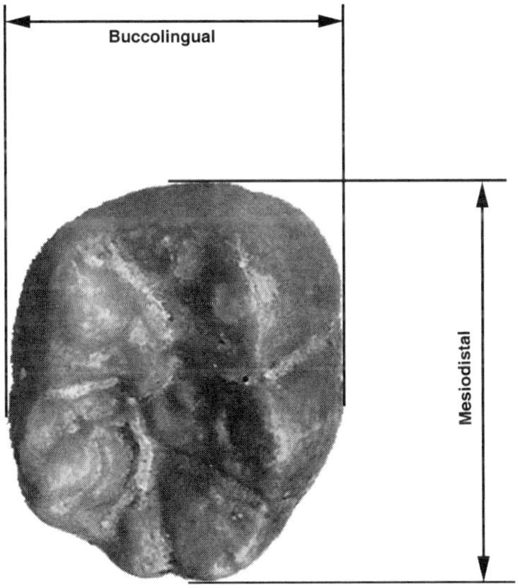

Figure 5.1. Mandibular molar with mesiodistal and buccolingual dimensions indicated.

Measurement Locations Are Variable

Because the mesiodistal and buccolingual dimensions are located at different locations on different teeth (Figure 5.2), the processes of growth disruption, postmortem taphonomic alterations, and modifications owing to oral health conditions are distributed throughout the dentition. For example, buccolingual dimensions of the incisors and canines are located near the cervico-enamel junction, whereas the mesiodistal dimensions for these teeth are located near the occlusal surface (incisors) or approximately at 75 percent of crown height as measured from the cervico-enamel junction (canines).

For premolars, mesiodistal and buccolingual dimensions are located between 50 and 75 percent of crown height, and for molars both measurements are located around 50 percent of crown height. This variation in measurement location is beneficial for two reasons. First, because teeth do not remodel, some measurements are more or less prone to growth perturbation, which provides a comparative method for evaluating ontogenetic plasticity. Second, the different crown positions of measurements mitigate the effects of adverse taphonomic or disease processes on missing data patterns and prevalence (discussed below).

Data Collection Protocol

My data collection strategy must consider the competing interests of resource management (time) and maximization of genotypic coverage. Ignoring time constraints, it might be preferable to measure all dental dimensions for every

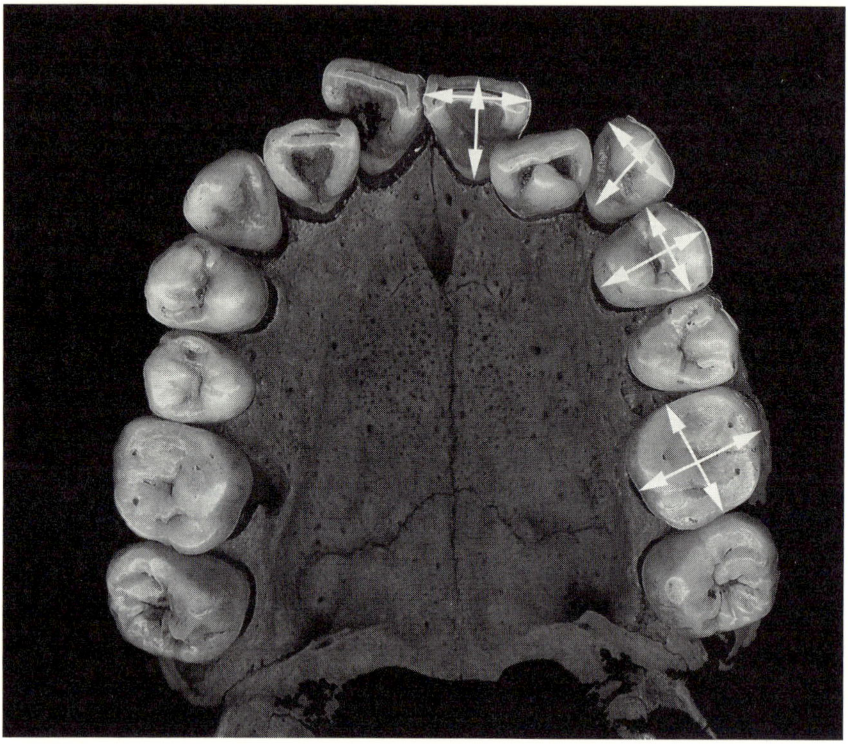

Figure 5.2. Adult maxilla demonstrating the specific teeth measured in this analysis and the mesiodistal and buccolingual dimensions for each tooth. Mandibular dimensions differ only in the inclusion of the second rather than the first incisor.

individual surveyed. However, because 900 individuals are included in this study, and each individual has a maximum number of 32 teeth (each with a length and width dimension), there were 57,600 possible measurements to record. Time constraints notwithstanding, previous research suggests that tooth size represents only a handful of underlying genetic loci, and therefore exclusion of some measurements will not significantly impact my ability to evaluate the research hypotheses. For example, Potter et al. (1968) identified three genetic tooth size components (mesiodistal and buccolingual posterior tooth size, mesiodistal anterior tooth size, and buccolingual anterior tooth size), and Potter and colleagues (1976) recorded eleven independent genetic components (four maxillary, seven mandibular); Lombardi (1975) identified five factors (molar size, premolar size, mesiodistal anterior tooth size, buccolingual anterior tooth size, and a combined mesiodistal/buccolingual incisor component); and Townsend and Brown (1979) identified four primary factors (molar size, mesiodistal anterior tooth size, buccolingual anterior tooth size, and a buccolingual premolar component). Therefore, the 64 possible measurements per individual represent between 3 and 11 independent genetic factors and can, therefore, be significantly reduced for logistical reasons.

This begs the question of which measurements can be excluded without significant loss of resolution. One finding common to dental inheritance studies is that extracted genetic factors do not cross tooth types; often all members of a tooth class (incisor, canine, premolar, molar) share the same underlying genetic factor of size expression. In addition, dental dimensions exhibit a high degree of intercorrelation, particularly among members of the same tooth class (Filipsson and Goldson 1963; Garn et al. 1965b, c; Garn, Lewis, and Wallenga 1968; Goose 1963; Kieser 1990; Moorrees and Reed 1964; Potter et al. 1976; Schnutenhaus and Rösing 1998). I can therefore measure only one member of a tooth class for the maxillary and mandibular dental arcades without sacrificing genotypic representation. Previous research indicates that the mesial-most teeth (also called the polar or key teeth) are preferable in this regard.

The tendency of polar teeth (I^1, I_2, C^1, C_1, P^1, P_1, M^1, M_1) to exhibit less variability is well documented (Dahlberg, 1951; Garn et al. 1965c, 1967b, c; Garn, Lewis, and Wallenga 1968), suggesting greater evolutionary stability and higher narrow-sense heritabilities resulting from diminished environmental and ontogenetic variation. As early as 1937, Newton reported increased second molar size variability (in comparison to the first molar) for triplets. Alvesalo and Tigerstedt (1974) reported higher heritabilities for the maxillary central incisor, canine, and first premolars and for the mandibular second incisor, first premolar, and first molars (hence my inclusion of I2 rather than I1; see also Kieser 1990). Dempsey et al. (1995) documented heritability differences of two to three percentage points between the first and second incisors, again suggesting that distal members of a tooth class demonstrate decreased heritability. This often results in a higher percentage of agenesis for distal tooth class members, particularly for the upper lateral incisors and third molars (Alvesalo and Tigerstedt 1974:315).

Inferences based on varied data sources and research projects suggest that dental size dimensions based on distal members of a tooth class are subject to greater environmental variation and agenesis resulting in smaller intercorrelations and lower realized narrow-sense heritabilities. This finding indicates that tooth sizes of polar teeth provide the best source of data for evolutionary genetic research and that inclusion of distal teeth is largely repetitive and incorporates a considerable degree of unnecessary environmental variation in the data set. Therefore, in addition to the time burden imposed by complete dental data recovery, inclusion of distal teeth may actually decrease my ability to investigate evolutionary genetic research questions.

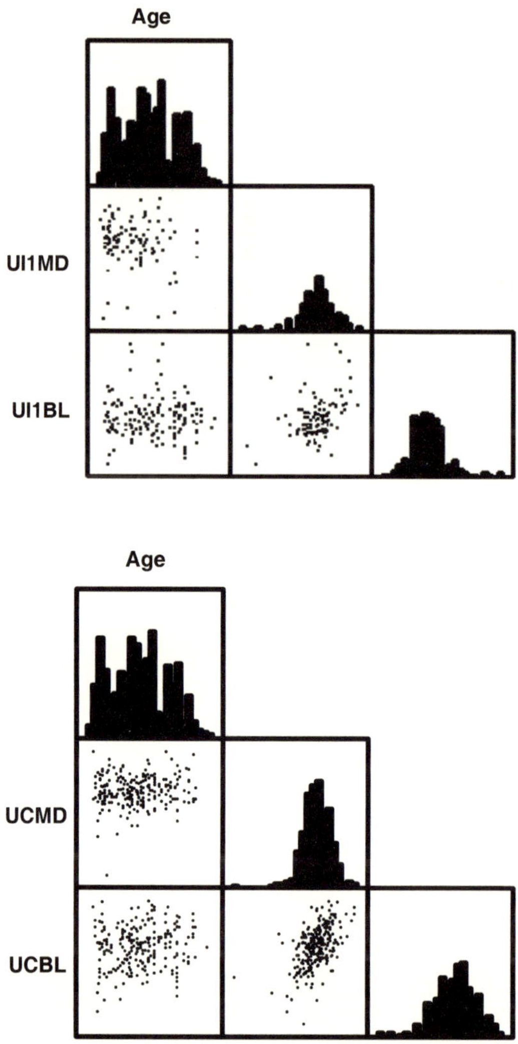

Figure 5.3a–d. Scatterplot matrix assessing the correlation between maxillary tooth size and age for the incisor (a), canine (b), premolar (c), and molar (d).

Measurement Protocol

Dental measurements were recorded to the nearest tenth of a millimeter using vernier sliding calipers. For the maxillary dentition, data for central incisors (I^1), canines (C), first premolars (P^1), and first molars (M^1) were collected. For the mandibular dentition, data for the lateral incisors (I_2), canines (C), first premolars (P_1), and first molars (M1) were collected. Crown dimensions were defined and measured as maximum values (Figure 5.3a–d) (see Kieser 1990;

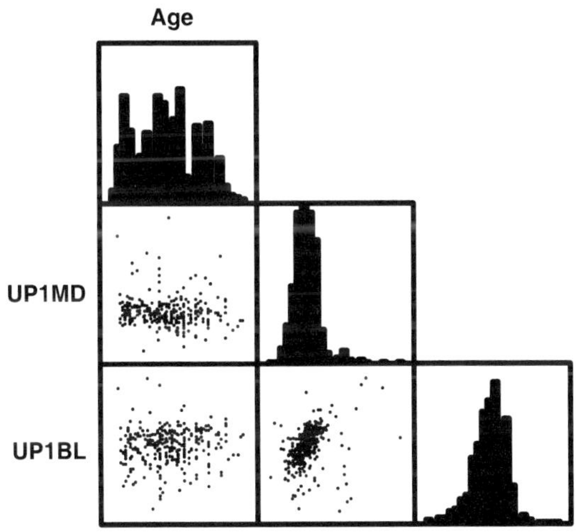

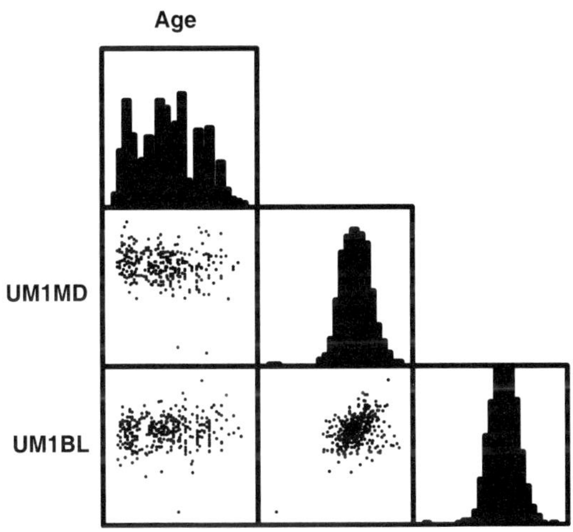

Figure 5.3 *Continued*

Mayhall 1992; Moorrees 1957) for compatibility with published comparative data and with the repatriated Santa Catalina de Guale de Amelia sample (Larsen, personal communication, 2000). Leftside measurements were recorded; however, antimere substitution was used to bolster limited sample sizes. While this may create problems for morphological features (Scott and Turner 1997), previous research has demonstrated highly significant correlations between deciduous and adult antimeres (Moorrees and Reed 1964; Schnutenhaus and Rösing 1998) or no statistically significant difference between right and left teeth within the same individual (Dempsey et al. 1995; Garn et al. 1967a;

Lundström 1967; Potter et al. 1976; Zilberman et al. 1995), suggesting that left and right tooth morphologies are controlled by homologous genes.

Factors that may affect genetic tooth form were identified prior to data collection and were subsequently used to exclude specific observations from the analysis (Beyer-Olsen and Alexandersen 1995; Goose 1963; Grant et al. 1979; Kieser 1990; Kolakowski and Bailit 1981). The following measurements were excluded: (1) those affected by macroscopic enamel hypoplasias, most often buccolingual canine dimensions; (2) those affected by large interproximal caries which may decrease mesiodistal or buccolingual dimensions; (3) those affected by excessive gingival calculus deposits, most often buccolingual dimensions of the anterior dentition; (4) those affected by interstitial attrition, most often mesiodistal dimensions of the incisors and molars; and (5) those affected by occlusal attrition, least often buccolingual diameters of the incisors, canines, and premolars.

Preliminary Analyses

Several pre-analysis data treatments were implemented to mitigate the influence of nongenetic factors in determining patterns of phenotypic variation within the mission populations. Despite intentional exclusion of measurements obviously affected by dental attrition and pathology, it is possible that non-macroscopically diagnosable alterations owing to attrition were missed. If so, samples primarily composed of younger individuals should exhibit less variability and vice versa. Evaluation and subsequent removal of age effects assures that variance differences are the result of population genetics and not the age structure of the samples. Intra-observer error results from differences in measurement technique throughout the course of data collection. Continued refinement and improvement of accurate recording methodologies may increase sample-specific variation for those samples analyzed earlier in the data collection process and vice versa. Finally, inter-observer error results from differences in measurement technique between observers and becomes problematic when data from multiple sources are compared in a comprehensive research project. Because of repatriation of some samples before initiation of this project, pooling of data collected by two different observers was required.

Age Effects

Minor decreases in crown dimensions are difficult to identify with the naked eye. However, it is clear that some dimensions are particularly prone to attrition-related loss. Kieser (1990) analyzed data from van Reenen's (1961, 1964) San Bushmen study on dental attrition and found the first incisors and first molars most susceptible to mesiodistal size reduction. Canines and second molars were the least susceptible. My personal observations of the mission materials produced similar results. Interstitial wear was most evident on the mesiodistal incisor and molar dimensions and to a lesser extent the premolar and canine dimensions. I would estimate that mesiodistal maxillary incisor dimensions were compromised by the age of 12. Owing to their low position on the crown, buccolingual dimensions of the incisors and canines, and to a lesser extent the premolars, were relatively unaffected by attrition (but calculus was a problem—see above). The amount of dimensional reduction can vary between 0.9 and 2 mm (Kieser, 1990), leading some to conclude that only unworn teeth should be used in dental studies (Kieser 1985; Kieser et al. 1985a, b, c), whereas others have advocated an estimation procedure for the inclusion of moderately worn teeth (Doran and Freedman 1974). I adopt the former approach.

I used two simple techniques to evaluate the data set for age dependencies. I first generated bivariate plots of individual crown dimensions and estimated age-at-death. This allows me to assess visually the relationship between age and tooth size, to evaluate trend directions in the data set, and to determine whether the variable relationships are generally linear. Pearson correlation coefficients supplement visual representations and provide enumerated estimates of the linear relationship between crown size and age. If interstitial and occlusal attrition significantly reduced crown size, I expect negative correlations between age and tooth size; that is, increasing age is associated with decreasing crown size.

Scatterplot matrices based on the complete data set are presented in Figures 5.3a–d and 5.4a–d for the maxillary and mandibular arcades. Overall, maxillary dimensions (with the exception of UP1MD) demonstrate little evidence for a positive or negative trend, and the data appear to be randomly distributed in a horizontal data band. Mandibular incisor and canine dimensions also appear to be unrelated to age. However, mandibular premolar and molar buccolingual dimensions appear to demonstrate a positive trend, while mesiodistal molar dimensions demonstrate a negative trend. There is little evidence for nonlinearity in any of these plots.

The aggregate sample correlation and regression coefficients are presented in Table 5.2. Twelve of 16 correlation coefficients are negative, suggesting a decreasing average tooth size with increasing individual age. Only one correlation (UM1MD), however, is statistically different from 0 at the 5 percent alpha level, a robust result given family-wise error expectations and the robust sample size for most comparisons (which reduces Type II error).

Although using the aggregate sample of measurements is beneficial for increasing sample size and statistical power, doing so assumes that the age distribution of the samples included in the aggregate data set are equal between sites and that tooth size is independent of the sample (unlikely). For example, a significant correlation could be observed if one sample contained young individuals with genetically larger teeth on average and a second sample contained older individuals with genetically smaller teeth on average. In this situa-

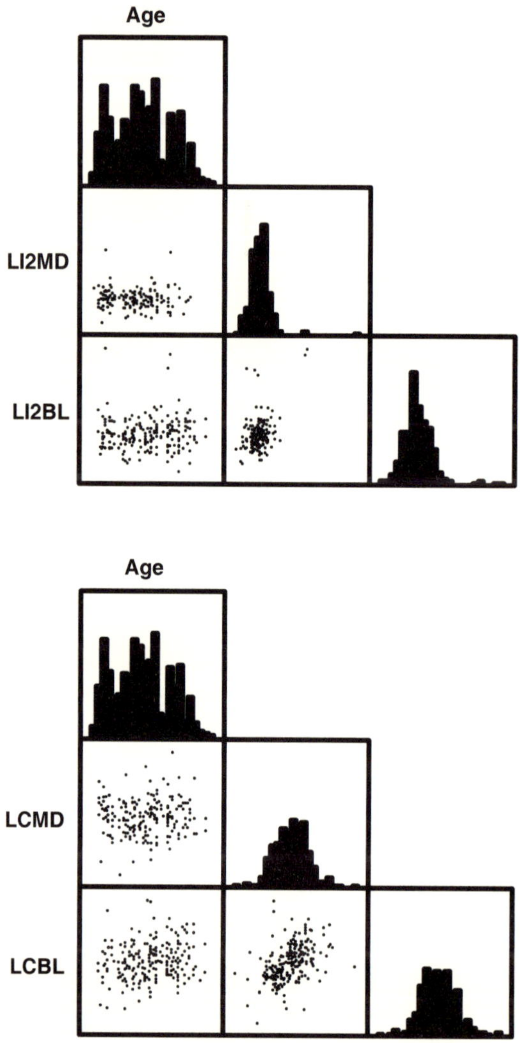

Figure 5.4a–d. Scatterplot matrix assessing the correlation between mandibular tooth size and age for the incisor (a), canine (b), premolar (c), and molar (d).

tion, there would be an apparent decrease in tooth size with age; however, this reflects population genetics and not the degree of mechanical wear. To avoid this problem, I also consider tooth/age correlations for individual samples (with sample sizes larger than 10). These data are presented for the maxilla in Table 5.3 and for the mandible in Table 5.4.

Of the samples large enough to be included, four had at least one significant correlation (Irene Mortuary—UI1BL, LP1MD; San Luis—UM1BL, LM1BL;

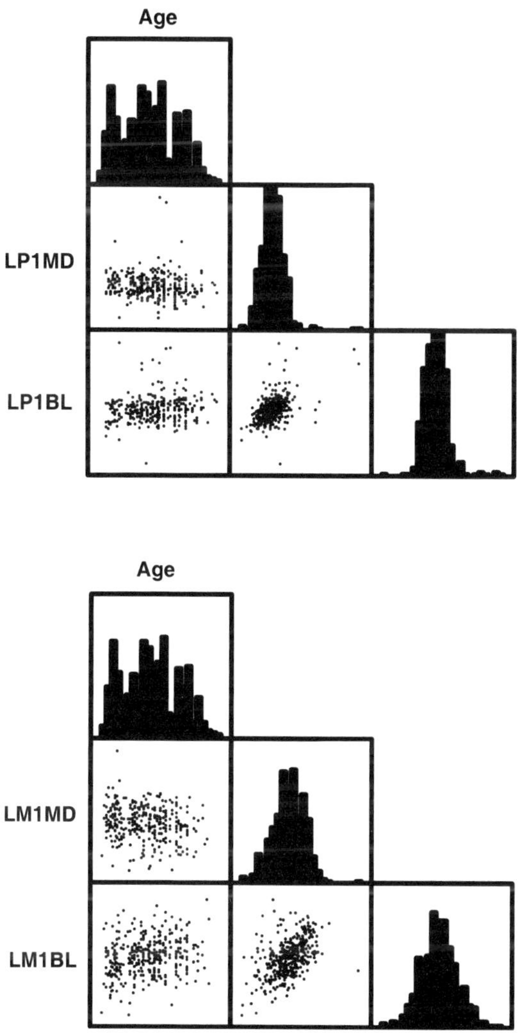

Figure 5.4 *Continued*

Santa Catalina de Guale—UM1MD, LP1BL, LM1MD; Santa Catalina de Amelia—LP1MD, LM1BL). However, if assessing crown loss with age is the hypothesis of interest, then only significant negative correlations should be considered. Therefore, based on the results presented in Table 5.2, UI1MD will be excluded from further consideration and all significant negative correlations in Tables 5.3 and 5.4 will be excluded from univariate tests that include that particular sample (Irene Mortuary—UI1BL; Santa Catalina de Guale—UM1MD, LM1MD; Santa Catalina de Guale de Amelia—LP1MD).

Table 5.2. Correlation between tooth size and age-at-death

Tooth	r(with age)	r (p-value)	n
UI1MD	–.593	.009[a]	128
UI1BL	–.048	.849	190
UCMD	–.011	.967	290
UCBL	–.128	.612	290
UP1MD	–.172	.494	372
UP1BL	.165	.513	386
UM1MD	–.074	.772	396
UM1BL	–.191	.447	439
LI2MD	–.435	.087	196
LI2BL	–.129	.597	239
LCMD	.229	.362	226
LCBL	–.091	.718	234
LP1MD	.285	.251	383
LP1BL	.301	.225	389
LM1MD	–.455	.058	393
LM1BL	–.067	.791	424

[a] significant at p=.05 level.

Table 5.3. Site-specific correlations between tooth size and age for the maxilla

	Correlation (p-value, sample size)			
Site	I1MD	I1BL	CMD	CBL
Irene Mortuary	—	–.55(.03, 14)	.23(.45, 13)	–.03(.89, 17)
Irene Mound	.15(.51, 19)	.04(.79, 41)	.16(.27, 46)	.23(.08, 52)
San Luis	—	—	–.09(.71, 17)	—
SCDG	–.19(.15, 54)	–.07(.70, 32)	.08(.39, 95)	.17(.09, 91)
SCDG-Amelia	—	.06(.65, 49)	–.04(.75, 44)	.04(.76, 46)

Site	P1MD	P1BL	M1MD	M1BL
Irene Mortuary	.68(.00, 14)	.25(.35, 15)	–.02(.43,21)	.03(.88,26)
Irene Mound	–.07(.58, 51)	–.08(.52, 56)	.03(.82,52)	.10(.38,74)
San Luis	–.26(.20, 24)	.03(.86, 26)	–.12(.53,29)	.37(.03,31)
SCDG	–.10(.25, 129)	.05(.51, 129)	–.18(.03,137)	.07(.38,136)
SCDG-Amelia	–.22(.11, 48)	.11(.43, 48)	–.23(.11,47)	.17(.22,51)

Note: SCDG = Santa Catalina de Guale.

Table 5.4. Site-specific correlations between tooth size and age for the mandible

	Correlation(p-value, sample size)			
Site	I2MD	I2BL	CMD	CBL
Irene Mortuary	.12(.71, 11)	.18(.46, 17)	.33(.19, 17)	.12(.56, 24)
Irene Mound	.22(.17, 38)	–.03(.82, 56)	.04(.78, 40)	.13(.33, 54)
San Luis	–.33(.34, 10)	—	–.28(.37, 12)	—
SCDG	–.00(.98, 78)	–.03(.77, 59)	.19(.14, 69)	.17(.17, 60)
SCDG-Amelia	—	.27(.05, 51)	–.06(.66, 48)	.18(.18, 55)
	P1MD	P1BL	M1MD	M1BL
Irene Mortuary	–.11(.63, 21)	.06(.76, 25)	–.05(.80, 25)	.30(.12, 27)
Irene Mound	–.25(.06, 57)	–.19(.12, 62)	.04(.76, 58)	.20(.09, 69)
San Luis	.18(.35, 29)	.24(.22, 26)	–.06(.72, 32)	.37(.03, 35)
SCDG	.09(.29, 131)	.22(.00, 132)	–.36(.00, 42)	.03(.71, 140)
SCDG-Amelia	–.39(.00, 50)	.09(.48, 52)	–.20(.21, 40)	.32(.03, 44)

Note: SCDG = Santa Catalina de Guale.

Intra- and Inter-Observer Error

Measurement error consists of both inter-observer error, error resulting from differences in the way two observers define and record identical measurements, and intra-observer error, error resulting from inconsistent data recording methods of a single observer (Kieser 1990; Schnutenhaus and Rösing 1998). While the cause of the former is easy to appreciate, the latter may result from myriad factors, including choice of calipers, caliper calibration, lighting or lab conditions, observer fatigue, and, perhaps most significantly, observer practice. Some of these are easy to control (for example, using the same set of calipers which are calibrated frequently), while others are more difficult to manage actively (fatigue is part and parcel of the data collection experience).

To evaluate levels of intra-observer error, I remeasured a subsample of teeth from six different samples. The first set of measurements was collected in February 2001, the second set in May 2001. The average difference between measurement sets 1 and 2, pooling all variables, was .01 mm with a 95 percent confidence interval ranging from –.01 to .12. In addition, statistical tests indicated little systematic error or inconsistency in my data collection technique. I used a one-sample t-test to evaluate the hypothesis that the mean difference between measurements sets 1 and 2 (for all variables) was equal to 0. The p-value was not significant ($t = 0.178$, $df = 227$, $p = .861$). Considering my primary interest in phenotypic variability, I also assessed whether the two measurement sets differed in statistical variance measures. An F-test indicated no

Table 5.5. Summary statistics and p-values for intra-observer error analyses

Measure	Mean diff	P-value T-test	Variance 1	Variance 2	P-value F-test
UI1MD	.030	.384	.584	.596	.474
UI1BL	.050	.449	.559	.560	.997
UCMD	.057	.402	.584	.604	.451
UCBL	.028	.087	.821	.817	.496
UP1MD	.049	.379	.783	.803	.457
UP1BL	.198	.280	.758	.735	.443
UM1MD	.045	.645	.785	.788	.489
UM1BL	.031	.210	.779	.780	.495
LI2MD	.036	.653	.162	.165	.478
LI2BL	.030	.999	.903	.900	.496
LCMD	.027	.514	.856	.873	.487
LCBL	.007	.500	.502	.495	—
LP1MD	.029	.040	.833	.853	.463
LP1BL	.051	.537	.853	.845	.487
LM1MD	.043	.453	.736	.713	.456
LM1BL	.051	.537	.853	.845	.487

Note: The sample size for LCBL was insufficient for this test.

systematic difference in variability (F = 1.00, p = .475, df = 227, 227). In Table 5.5, I present results for each measurement separately. There is little evidence for intra-observer error in these data, and the single significant test (LP1MD, p = .04) can be attributed to family-wise error.

Unfortunately, I was unable to collect data for two critical samples for this project: Santa Catalina de Guale de Amelia had been repatriated, and San Luis de Talimali was unavailable for analysis. Data for both samples were generously provided by Clark S. Larsen. For all variables combined, the average difference between measurement sets was .01 mm with a 95 percent confidence interval ranging from –.009 to .04. A one-sample t-test evaluating whether the average difference between measurement sets equaled 0 was not significant (t = 1.25, df = 140, p = .110). Differences in recorded variability were also not significant between measurement sets (F = 1.02, df = 140, 140, p = .440). Tooth-specific data are presented in Table 5.6. There is little evidence for systematic collection error for any variable. Although a single p-value comparing mean differences and a single p-value comparing variability differences reached statistical significance, neither of these tests remains significant when alpha levels are adjusted for family-wise error.

Table 5.6. Summary statistics and p-values for inter-observer error analyses

Measure	Mean diff	P-value T-test	Variance 1	Variance 2	P-value F-test
UI1BL	.032	.707	.164	.323	.030[a]
UCMD	.109	.122	.141	.213	.200
UCBL	.006	.696	.184	.206	.400
UP1MD	.074	.082	.200	.186	.440
UP1BL	.020	.646	.256	.289	.410
UM1MD	.180	.070	.187	.223	.360
UM1BL	.016	.282	.308	.304	.490
LI2BL	.050	.423	.166	.213	.200
LCMD	.017	.768	.209	.151	.200
LCBL	.014	.609	.177	.157	.400
LP1MD	.058	.082	.162	.199	.200
LP1BL	.048	.090	.181	.200	.300
LM1MD	.087	.016*	.122	.133	.400
LM1BL	.035	.211	.150	.212	.100

Note: Larsen did not collect mesiodistal dimensions for the incisors, therefore these data were unavailable for comparison.

[a] alpha protection adjusts p-value to .004 for a test at the 5% level.

Statistical Analysis of the Data

Several univariate and multivariate statistical tests were used to evaluate the research hypotheses of phenotypic variance changes. Statistical details are presented in Appendix A for those interested, and only the most general details are presented here. For univariate comparisons, three tests were used. An F-test was used to compare pair-wise combinations of samples (two groups only), and Bartlett's test was used to compare variances for a single variable for three or more groups. Neither of these tests is optimal, given their sensitivity to sample normality (Sokal and Rohlf 1995). A robust alternative to Bartlett's test for multigroup comparisons is Levene's test based on sample medians. Levene's test is more conservative than Bartlett's test and maintains the specified alpha level for any continuous distribution regardless of distributional shape. In combination, then, Bartlett's and Levene's tests offer two vastly different approaches to assessing univariate variance differences in two or more samples. Both are preferable to multiple two-sample F-tests because the familywise error rate is minimized.

In addition to formally testing differences in sample variability, I also calculated Bonferroni confidence intervals for the individual sample standard deviations. Bonferroni confidence intervals differ from standard confidence intervals in that they divide the family-wise error rate among the confidence intervals being simultaneously calculated, thus ensuring that the 5 percent error rate is maintained over all intervals. For a series of six 95 percent confidence intervals, the individual confidence intervals have a .99 level because the Type I error rate is divided among the six intervals. This creates longer than normal sample intervals, but any differences observed are robust for this reason. In addition, because the confidence intervals are based on the chi-square distribution, they are asymmetric around the sample standard deviation.

As a more liberal alternative, I also generated confidence intervals for the ratio of variances for each between-sample comparison. If the population variances are in fact equal, then the confidence interval for the ratio of two variances should include 1. Because I calculated all ratios using the larger variance as the numerator, all ratios are greater than 1 and therefore only the lower bound is pertinent to this analysis. If the range between the lower confidence interval bound and the observed ratio covers 1, then the two sample variances are not considered statistically different.

Despite weaknesses in using univariate statistical techniques, these approaches avoid problems associated with missing data imputation. Multivariate analyses, on the other hand, usually require complete matrices, necessitating that missing data be estimated prior to implementation of variance analyses. Nevertheless, a multivariate approach may identify differences in covariation among variables not observed in the univariate results (minor differences may exponentiate in multivariate space) and is therefore preferable in this regard. I use two multivariate methods in this analysis, Van Valen's test and determinant ratio analysis.

Van Valen's test is based on the distance of each observation from the within-group mean. The benefit of this analysis is that data imputation is unnecessary because the test is independent of the variance covariance matrix (as opposed to Levene's multivariate generalization). Observations are first standardized to mean 0 with unit variance across groups. Then, site-specific means are computed and used to estimate the average difference between the standardized within-group observations and the within-group mean. The resulting values are compared between groups using a Student's t-test or ANOVA, depending on the number of groups being compared. The theory behind the test is quite simple. In samples where distances from the within-group mean are large, the summary test statistic will also be large. The parametric test of the distance values determines whether one sample has a larger average summary value than another sample. One limitation of Van Valen's test is the assumption that a majority of the variables included in the test are more variable than the other samples; otherwise, the distances tend to cancel one another. On the other hand, the additive nature of the test statistic acts to highlight minor variance differences that might not be significant in isolation.

To implement determinant ratio tests (see Green 1976), missing data were first imputed using the MISSING module available in SYSTAT v 10.0 (Wilkinson et al. 1996). Determinant ratio analysis was then performed using the RANDET2 Fortran program written by Lyle Konigsberg. The program first centers each group to mean 0 and calculates the natural logarithm of the ratio of determinants for the two samples. If two determinants are equal, the determinant ratio equals 1. Because ln(1) = 0 and ln(.01–.99) is negative, a negative logarithmic determinant ratio indicates that the denominator sample is more variable than the numerator. The converse is indicated by logarithmic determinant ratios greater than 1. The RANDET2 program assesses statistical significance using a bootstrapping resampling procedure (see Edgington 1987; Konigsberg 1988; Petersen 2000). For two samples the data matrices are combined into a single matrix. The rows are then randomly shuffled; a subset of cases equal to the original group 1 sample size is designated, and the remaining cases are assigned to group 2. The determinant of the variance-covariance matrix is calculated for each sample and recorded. The procedure is then repeated 999 times until a distribution of determinant ratios is produced. The exact p-value for the test is provided by the proportion of permuted determinant ratios greater than or equal to the actual observed determinant ratio (Konigsberg 1988:478; Petersen 2000).

The original dissertation on which this book is based included a lengthy development of a new methodology specifically for use on the mission data set (Stojanowski 2001). This work has subsequently been revised and published in Stojanowski (2003). Because of limitations of the analysis, it can only be applied very narrowly; only the Guale skeletal samples are large enough and complete enough for consideration.

The problem posed in Stojanowski (2001, 2003) is a simple one. When transitioning from a predominantly lineage-based burial program to an aggregate, cemetery-oriented burial program, one expects greater variability in the latter because a larger percentage of the overall population is buried in the aggregate cemetery. This simple fact influenced the decision to use aggregate precontact samples for both Apalachee and Guale. Irene Mound and Mortuary, however, are large samples, likely aggregate, and possibly including burial catchments equivalent to the mission period cemeteries. Herein lies the second problem. Given the documented degree of in-migration and population resettlement during the mission period, doctrinas could very well have been a diverse polyethnic community but with limited interbreeding between these segments of the population. Further, given population turnover rates, it is possible that epidemics left missions empty and the Spanish relocated populations to fill these important locations within the mission chain. If these populations buried their dead in the same cemetery, then the burial sample would appear variable statistically owing to the confounded effects of genetic admixture and population aggregation. The matrix decomposition model of Stojanowski (2001, 2003) tests this assumption by generating estimates of the likely number of distinct burial populations interred within a single cemetery.

The mechanics of the matrix decomposition model are relatively straightforward. Inter-individual genetic and burial distance matrices are generated for each sample. If there is a significant correlation between these matrices, it suggests that burial placement was concordant with kinship relationships within the cemetery for a *single* population. Closer burial proximity is associated with smaller genetic distances and vice versa. However, if multiple lineages or populations were buried within a single cemetery but generally used discrete interment areas, then the overall correlation between grave and genetic distances is not expected to be significant because all interpopulation comparisons should be random with respect to this relationship.

The matrix decomposition analysis entails several steps (see Stojanowski 2003 for complete derivation).

(1) Estimate the correlation between genetic and burial distance matrices.
(2) If insignificant, subtract the genetic distance between two graves from the burial distance between these graves. Repeat for all graves. If the result is negative, it suggests that the pair of individuals is less related than expected under the assumption of kin-structured burial and likely represents a comparison of individuals from discrete burial subpopulations.
(3) The proportion of negative residuals decreases as the number of burial subgroupings in a cemetery increases.
(4) Statistical significance is assessed using confidence intervals for the difference in two proportions.

Dental measurement data provide a number of advantages over other classes of phenotypic osteodental data typically used in evolutionary research designs: good preservation, ease of measurement, statistical tractability, lack of ontogenetic plasticity, and variable measurement locations, all of which mitigate potential environmental and postdepositional noise.

Maximum mesiodistal and buccolingual dental diameters were collected for the maxillary central incisor, canine, first premolar, and first molar, and for the mandibular lateral incisor, canine, first premolar, and first molar. Measurements were recorded to the nearest tenth of a millimeter using vernier sliding calipers. Measurements visibly affected by dental attrition, calculus, caries, or hypoplastic defects were excluded. Age effects were evaluated using correlation analysis, resulting in the omission of upper central incisor mesiodistal diameters outright, in addition to several sample-specific measurements that demonstrated statistically significant negative correlations between age-at-death and tooth size. Inter- and intra-observer errors were evaluated and found to be minimal.

Analysis of variability differences among archaeological skeletal samples utilized univariate and multivariate statistical tests. The former consisted of F, Bartlett's, and Levene's tests. Pairwise differences were evaluated using sample-specific Bonferroni confidence intervals and confidence intervals for variance ratios. Multivariate analyses consisted of Van Valen's test and determinant ratio analysis. A test designed to parcel the effects of admixture from population aggregation was implemented for the Guale samples.

Note

1. I must confess I am still challenged to accurately and consistently record labial curvature on maxillary incisors and the degree of double shoveling on maxillary premolars and canines.

Part III
The Synthesis

6 Demographic Transformations among the Apalachee

As detailed in chapters 2 and 3, the Apalachee were generally distinct from their neighbors in terms of language, culture, subsistence strategy, settlement pattern, and political organization. They also likely weathered the rigors of the contact period better than their contemporaries, which influences hypothesized microevolutionary changes for the province.

Based on my reading of the historical literature and the research model generated in previous chapters, I propose the following for Apalachee Province. Both Stage 1 (local aggregation to the doctrina) and Stage 2 (the consolidation of doctrinas) congregación would have had limited effect on variability levels *within* the province. First, these efforts at reorganization were not used extensively (Hann 1986b:387), if at all, because population sizes remained comparatively stable in Apalachee throughout much of the mission period (Hann 1988). Second, available data suggest that the Apalachee spoke the same language, were culturally and politically unified, and were not internally divided to any significant extent (see Hann 1988; Jones and Shapiro 1990; Scarry 1992:17). Social organizational principles indicating village exogamy and an exogamous and geographically dispersed clan structure suggest that mate exchange was common *between* local villages and doctrinas (Bushnell 1978b), and related elites were distributed widely throughout the province. All villages were united by complex ties of marriage.

Patterns of regional linguistic and political structure, combined with Apalachee warfare practices, suggest biological circumscription of the territory. Therefore, both Type 1 (Christian in-migration) and Type 2 (nonlocal in-migration) reducción *would* have significantly increased variability levels among the Apalachee and would have been particularly important after the rebellions in the western provinces (in 1647 and 1656). I propose a significant increase in interethnic contact and intermating between Apalachee and non-Apalachee immigrants at this time. Because Apalachee was the seat of the

Table 6.1. Apalachee Province skeletal samples

Sample	Size	Period	Source
San Pedro y San Pablo de Patale	58	Early Mission	Jones et al. 1991
San Luis de Talimali	75	Late Mission	Shapiro and Vernon 1992
Lake Jackson	23	Late Precontact	Jones 1982
Waddell's Mill Pond	1	Late Precontact	Gardner 1966
Killearn Borrow Pit	2	Late Precontact	Fla. State Site File
Snow Beach	6	Late Precontact	Magoon et al. 2001

regional Hispanic governing community situated at mission San Luis, Spanish residents would have provided an additional source of increased genetic variability (McEwan 1991a, 2001). I therefore predict that variability decreased with the transition from the late precontact to early mission periods (1633–50) via the action of random genetic drift in populations of rapidly decreasing size. I then predict increased variation during the later phases of the Spanish colonial period (1650–1704) owing to admixture with diverse population groups immigrating to the province.

Bioarchaeological Resources

As indicated in Table 6.1 and Figure 6.1, two mission period cemeteries have been excavated, the pre-1650 San Pedro y San Pablo de Patale mission and the post-1656 San Luis de Talimali mission. That these burial populations represent nonoverlapping periods of interment provides a unique opportunity to examine microevolutionary processes and levels of phenotypic variation through two distinct time periods. However, the unique function and position of the San Luis population may conflate synchronic sociopolitical explanatory mechanisms with those owing to temporal factors.

San Pedro y San Pablo de Patale (8Le152) is a pre-1650 mission. The site was discovered in 1968 by B. Calvin Jones of the Florida Bureau of Archaeological Research and excavated over a four-year period between 1968 and 1972. Rochelle Marrinan of Florida State University has excavated the site intermittently over the past 15 years (Marrinan 1993). All human remains included in this analysis derive from the Jones excavations (see Jones et al. 1991).

Despite frequent mention of the Patale community in various contemporary historical documents, including the 1655 mission enumeration, the 1657 Rebolledo visitation record, the 1675 Florencia mission list, the 1675 Calderón enumeration, the 1677 Leturiondo visitation, the 1680 Cabrera mission list, the 1689 Bishop of Santiago de Cuba mission list, and the 1694–1695 Florencia

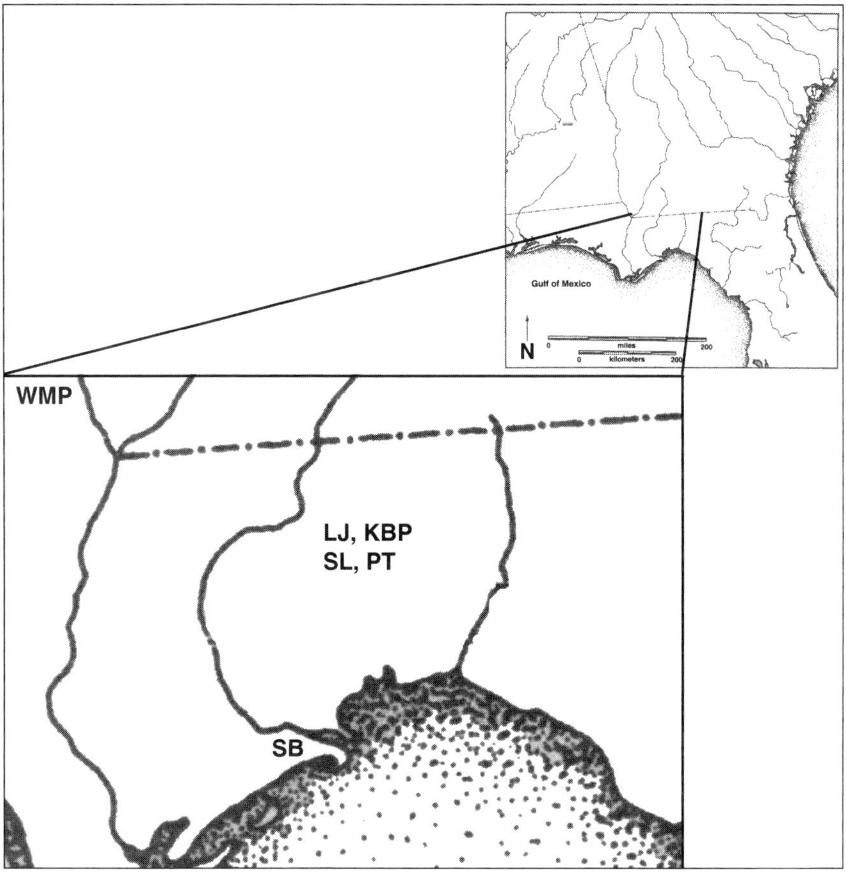

Figure 6.1. Map of Florida showing locations of the Apalachee samples used in this analysis. WMP = Waddell's Mill Pond, LJ = Lake Jackson, KBP = Killearn Borrow Pit, SL = San Luis, PT = Patale, SB = Snow Beach (modified from Larsen 2001: Fig. 2.1).

visitation (Jones et al. 1991), these sources are likely discussing secondary or perhaps tertiary locations of the doctrina serving the Patale community. The cemetery under consideration here (8Le152) probably predates 1650, based on majolica types recovered from the site which predate the mid-17th century (Columbia Plain Gunmetal, Mexico City White, San Luis Blue-on-White, and Fig Springs Polychrome), as well as those absent from the assemblage which postdate the mid-17th century (Puebla Polychrome, Abo Polychrome, and San Luis Polychrome; Jones et al. 1991:73). Given the *terminus post quem* of A.D. 1633 established by the onset of Franciscan proselytizing in Apalachee (Hann 1988), the Patale cemetery at site 8Le152 represents burial activity that occurred between 1633 and approximately 1650.

The Patale cemetery was located within the boundary of the church walls.

Jones uncovered 65 burial pits of which 58 produced preserved human tissue. The total sample size for the cemetery was 67 poorly preserved individuals. They were buried in an extended and supine position, with heads facing northeast, and in nine distinct burial rows that were divided by a center aisle. Analysis of the demographic structure of the cemetery produced weak location-specific demographic patterning. Subadults tended to be buried on the left side of the aisle (68.2 percent) and adults on the right (58.5 percent). Also, while females were evenly divided by side, most male burials were interred to the right of the aisle (Jones et al. 1991). Initially, Jones et al. (1991) noted that the adult burials were grouped near the rear of the church; however, further excavation and spatial analysis by Marrinan (1993) suggested that the altar was incorrectly positioned by Jones. If the altar was located on the opposite end of the nave as that presented in the Patale report, the adult grouping is actually located near the altar end of the church.

San Luis de Talimali was one of the most important of the Apalachee missions, serving as the capital of the province from 1656 until 1704. At its peak, some 1,400 Apalachee lived under the jurisdiction of San Luis's chief, and by the end of the 17th century several hundred Spaniards (including soldiers, friars, and civilians) lived at the mission (Boyd et al. 1951; Hann 1988; McEwan 1991a, b, 1992, 1993, 2000, 2001). Its key political and military position, both for the Spanish government and for the indigenous population of the province, led to an integration of Spanish and native cultures during this time period. Archaeological and historical research at San Luis has developed a social history of the community by examining social institutions, technology, faunal and floral remains, disease pathology, and architecture introduced on the landscape of Spanish Florida (Cordell 2002; Hann 1988; Hann and McEwan 1998; Larsen and Tung 2002; Larsen et al. 1996; McEwan 1991a, b, 1992, 1993, 2000, 2001; Reitz 1993; Ruhl 2000; Scarry 1993; Shepard 2003).

The San Luis cemetery was identified by Gary Shapiro during the broad-scale survey in the spring of 1987 (Shapiro 1987). Clark Spencer Larsen directed the excavation and study of skeletal remains, while Bonnie G. McEwan directed other aspects of the church investigations. Fieldwork in the cemetery was completed in 1997 (Larsen and Tung 2002; McEwan 2001). Study of the burial population resulted in the identification of a minimum of 210 individuals. The majority of burials in the San Luis cemetery conformed to the Catholic pattern of interment typical of the 17th century. Individuals were laid to rest on their backs in shallow burial pits with legs extended, heads oriented east, and feet facing the altar, with their hands folded across the chest. Of the 210 excavated skeletons, 7 were granted coffin burial, suggestive of a unique status within the community. All other individuals were interred in burial pits lacking evidence of coffin association. Individuals interred in coffins were grouped near the front of the church.

Late precontact samples are exceedingly rare from the Apalachee/Fort Walton region. The largest from Lake Jackson is an elite mound burial context, whereas the other three samples (Killearn Borrow Pit, Snow Beach, and Waddell's Mill Pond) are smaller samples consisting of few individuals.

Lake Jackson is a large ceremonial center consisting of multiple mounds and other ceremonial features (Jones 1982). All burials in this analysis derive from mound 3, which was located in the southernmost row of mounds at the site. Stratigraphic analysis indicated 12 distinct floors, each of which probably supported a single-walled rectangular structure for housing burials (Jones 1982). Twenty-four burials were recorded; however, only 15 were excavated and collected. Most were primary, single interments buried in shallow pits covered with split logs and containing a variety of high status grave goods. The remains were poorly preserved; they were primarily of advanced age and represented both males and females (Jones 1982).

Based on the size of the site and the quality of grave goods included with the interments, Lake Jackson served as the seat of a powerful Mississippian chiefdom. The burials, therefore, likely represent a small, elite segment of the Fort Walton population that inhabited the area prior to European contact. Carbon dating of stratigraphic levels within the mound indicates a 400-year period of use beginning with an earlier premound midden (Jones 1982). The earliest mound construction (Floor 12) dates to A.D. 1240 ±90 years. The last building phase (Floor 1) dates to A.D. 1475 ±85 years. Floor 1 contains the greatest number of burials (n = 9), followed by floors 8 and 9 (n = 3), floor 10 (n = 2), and floors 2, 3, and 11 (n = 1). This stratigraphic profile indicates that ceremonial activity was increasing in frequency before its abrupt abandonment shortly before initial European contact.

Killearn Borrow Pit and Waddell's Mill Pond derive from nonmound burial contexts. The former represents a Fort Walton period council house, which produced poorly preserved human remains (Jones et al. 1991; Scarry 1992). Although the field inventory indicated an MNI of six individuals, only two had preserved dental remains. The latter sample produced only a single adult specimen (Gardner 1966).

Snow Beach is a protohistoric mound and midden located near the Gulf Coast (Magoon et al. 2001:18). Although the site had multiple archaeological components, the burials are associated with the later occupation of the site as evidenced by the identification of European trade goods dating to the first half of the seventeenth century (Magoon et al. 2001:20). Of the eight individuals, only five had preserved dental remains despite the overall excellent skeletal preservation. Burials were single (with one exception), primary, extended interments that represented Fort Walton elite, similar to (though postdating) the Lake Jackson sample. Isotopic analyses indicate considerable dietary diversity for the Snow Beach remains that included more marine proteins in comparison to contemporary inland samples (Magoon et al. 2001: 23–25).

In combination, the four precontact skeletal samples are representative of both elite and nonelite Apalachee from various burial contexts. The sites may, therefore, be fairly representative of an aggregate precontact population. Because of the individually small size of the precontact samples, they were combined into an aggregate late precontact sample (hereafter LPC) for comparative purposes. While problems with such an approach are obvious, I feel they are unavoidable. In addition, if anything, the spatial-temporal aggregation artificially inflates the estimate of variability for the late precontact period, making affirmation of predicted declines in variability a robust result. The Apalachee database, therefore, consists of three samples: a precontact aggregate sample, the pre-1650 Patale mission, and the post-1650 San Luis mission.

Analytical Results

Univariate Variability Differences

Sample sizes and standard deviations are presented in Table 6.2. Univariate analyses of phenotypic variability changes produced a number of interesting patterns. Consideration only of the magnitude of variance changes resulted in the following results.

(1) Comparison of the LPC and Patale samples indicates an increase in variability from the precontact to early mission period for 8 of 16 measurements, a result that is expected based on chance and one that indicates little evidence for an increase or decrease in variability during the incipient phases of missionization.
(2) When the LPC standard deviations are compared against those of the late mission period San Luis sample, 11 of 12 available measurements demonstrate larger standard deviations in the San Luis sample, suggesting a temporal increase in phenotypic variability.
(3) Comparison of the early and late mission period samples produced results similar to those for the precontact San Luis comparisons. San Luis was more variable than Patale for 10 of 11 measurements and, therefore, demonstrates excess phenotypic variability in comparison to both temporally ancestral populations.

Inferential analysis for univariate variables was accomplished with Bartlett's, F, and Levene's tests. P-values are presented in Tables 6.3 and 6.4 for the maxillary and mandibular dentitions, respectively. Three statistically significant tests were recorded; however, the Bartlett's test p-values did not correspond with the confidence intervals for the standard deviations. In other words,

Table 6.2. Apalachee standard deviations and sample sizes

Measure	LPC	PT	SL
UI1BL	4(.53)	6(1.23)	—
UCMD	7(.25)	13(.57)	24(.57)
UCBL	6(1.02)	4(.99)	10(.56)
UP1MD	9(.33)	15(.34)	30(.75)
UP1BL	13(.60)	17(.61)	32(.94)
UM1MD	9(.68)	16(.48)	36(.63)
UM1BL	10(.49)	15(.68)	age effect
LI2MD	4(.63)	11(.40)	10(.50)
LI2BL	3(.34)	7(.29)	—
LCMD	3(.57)	4(.31)	15(.71)
LCBL	2(.38)	2(.16)	6(.73)
LP1MD	7(.39)	13(.31)	39(.53)
LP1BL	8(.46)	10(.32)	36(.72)
LM1MD	4(.69)	13(.97)	37(1.83)
LM1BL	4(.44)	14(.47)	age effect

Note: LPC = Aggregate Late Precontact; PT = Patale; SL = San Luis. "Age effect" indicates the data were excluded because of a significant correlation with age.

the Bartlett's test indicated a significant difference between groups, but all confidence intervals for the standard deviations overlapped. In such situations, it is possible that the two samples with the least amount of overlap are driving the results. For the three cases in which this occurs (UP1MD, LP1BL, LM1MD), the data indicate that San Luis was significantly more variable than Patale. Further analysis by group (LPC vs. Patale, LPC vs. San Luis, Patale vs. San Luis) using F-tests produced multiple significant results: San Luis was more variable than Patale for four measurements (UP1MD, LP1MD, LP1BL, and LM1MD) and was more variable than LPC for two measurements (UCMD, UP1MD). None of the comparisons involving LPC and Patale were significant. Bonferroni p-value adjustments reduced the number of significant tests; however, the pattern of variability and the limited statistical information suggest that genetic variability increased post-1650. Confidence interval bounds for variance ratios are presented in Tables 6.5 and 6.6 for the maxilla and mandible, respectively. Although generally considered more liberal than individual sample confidence intervals, the data indicate no significant differences in variability for all variables for all sample comparisons.

Table 6.3. Apalachee maxillary univariate variance tests

Measure	Period[a]	n	sd	95% CI (sd)	Bartlett Stat (p-value)	Levene Stat (p-value)
UI1BL	LPC	4	.54	.28–2.54	.19(.20)	.07(.80)
	PT	6	1.23	.72–3.52		
UCMD	LPC	7	.26	.15–.70	4.32(.12)	1.11(.34)
	PT	13	.57	.39–1.07		
	SL	24	.57	.42–.87		
UCBL	LPC	6	1.02	.58–3.18	2.58(.28)	1.13(.35)
	PT	4	.99	.50–5.41		
	SL	10	.56	.36–1.19		
UP1MD	LPC	9	.33	.21–.75	12.97(.00)	1.18(.32)
	PT	15	.35	.24–.61		
	SL	30	.75	.57–1.09		
UP1BL	LPC	13	.60	.40–1.12	5.34(.07)	1.20(.31)
	PT	17	.62	.43–1.04		
	SL	32	.95	.72–1.35		
UM1MD	LPC	9	.68	.42–1.54	1.72(.42)	.48(.62)
	PT	16	.48	.33–.82		
	SL	36	.63	.49–.88		
UM1BL	LPC	10	.49	.31–1.04	0.30(.42)	.43(.75)
	PT	5	.69	.47–1.21		

[a] LPC = Aggregate Late Precontact, PT = Patale, SL = San Luis.

Multivariate Variability Differences

Multivariate analyses produced results consistent with those of the univariate procedures. For the multivariate Van Valen's test, nine variables were included (UCMD, UCBL, UP1MD, UP1BL, UM1MD, UM1BL, LP1MD, LP1BL, LM1MD) for comparison of variability in Lake Jackson, Patale, and San Luis (sample sizes were too small for the other precontact samples). UI1MD and LM1BL were excluded because of age effects, and UI1BL, LI2MD, LCMD, LCBL were excluded because of a lack of data in the Lake Jackson sample. The results of the ANOVA on the Van Valen's test statistics are presented in Table 6.7; the p-value was highly significant (p = .008). Both Fisher and Bonferroni multiple comparisons procedures generated identical results: San Luis was significantly more variable than both Patale and Lake Jackson. Patale was more

Table 6.4. Apalachee mandibular univariate variance tests

Measure	Period[a]	n	sd	95%CI (sd)	Bartlett Stat (p-value)	Levene Stat (p-value)
LI2MD	LPC	4	.63	.32–3.43	1.03(.60)	.09(.91)
	PT	11	.40	.26–.82		
	SL	10	.50	.32–1.07		
LI2BL	LPC	3	.34	.16–3.02	1.35(.65)	.00(.94)
	PT	7	.29	.18–.73		
LCMD	LPC	3	.57	.26–6.27	2.02(.36)	.96(.40)
	PT	4	.32	.16–1.73		
	SL	15	.71	.49–1.26		
LCBL	LPC	2	.38	.14–36.49	1.85(.40)	.83(.47)
	PT	2	.16	.06–14.92		
	SL	6	.74	.42–2.29		
LP1MD	LPC	7	.39	.23–1.06	4.30(.11)	1.23(.30)
	PT	13	.32	.21–.59		
	SL	39	.53	.42–.73		
LP1BL	LPC	8	.46	.28–1.11	7.76(.02)	2.17(.12)
	PT	10	.32	.21–.69		
	SL	36	.73	.56–1.01		
LM1MD	LPC	4	.68	.35–3.73	9.54(.01)	.12(.89)
	PT	13	.87	.58–1.63		
	SL	37	1.83	1.42–2.53		
LM1BL	LPC	4	.44	.22–2.4	0.98(.64)	.34(.54)
	PT	14	.48	.32–.87		

[a] LPC = Aggregate Late Precontact, PT = Patale, SL = San Luis.

variable than Lake Jackson; however, the difference did not reach statistical significance.

Determinant ratio analysis was more limited in its inclusion of samples owing to the requirements of data imputation. Because Lake Jackson and the other precontact samples were small with very sparse data matrices, the imputation procedure was unable to estimate a nonsingular covariance matrix for these samples. For this reason, only Patale and San Luis could be included in this analysis. The determinant ratio analysis was computed for only three variables (UP1MD, UP1BL, LM1MD) for the two mission samples. The determinant

Table 6.5. Apalachee maxillary univariate variance ratio confidence intervals

Measure	Comparison[a]	Ratio	num d.f.	den d.f	CI 95%-
UI1BL	LPC vs PT	2.28	5	3	.15
UCMD	LPC vs PT	2.19	12	6	.18
	LPC vs SL	2.19	23	6	.42
	PT vs SL	1.00	23	12	.33
UCBL	LPC vs PT	1.03	5	3	.07
	LPC vs SL	1.82	5	9	.23
	PT vs SL	1.77	3	9	.35
UP1MD	LPC vs PT	1.06	14	8	.26
	LPC vs SL	2.27	29	8	.58
	PT vs SL	2.14	29	14	.78
UP1BL	PC vs PT	1.03	16	12	.33
	LPC vs SL	1.58	31	12	.53
	PT vs SL	1.53	31	16	.60
UM1MD	LPC vs PT	1.41	8	15	.45
	LPC vs SL	1.08	8	35	.50
	PT vs SL	1.31	35	15	.50
UM1BL	LPC vs PT	1.41	4	9	.30

Note: Ratio = variance ratio for groups being compared, num d.f. = numerator degrees of freedom, den d.f. = denominator degrees of freedom, CI 95%- = lower bound of 95% confidence interval.

[a] LPC = Aggregate Late Precontact, PT = Patale, SL = San Luis.

ratio (ln det2/det1) was equal to –6.46 with a significant p-value (p = .031), indicating that San Luis was significantly more variable than Patale.

Processes Internal to Apalachee Province

Incipient Colonization

This analysis has demonstrated that levels of phenotypic variability did not change with the transition to the mission period. Three interpretations are possible: (1) The late precontact sample underestimates population genetic variability in the late precontact Apalachee, and there was no observed decrease in the Patale sample for this reason; (2) population size did not decline in Apalachee, at least to the extent that genetic drift was a significant evolutionary

Table 6.6. Apalachee mandibular univariate variance ratio confidence intervals

Measure	Comparison[a]	Ratio	num d.f.	den d.f.	CI 95%-
LI2MD	LPC vs PT	1.58	3	10	.33
	LPC vs SL	1.20	3	9	.24
	PT vs SL	1.25	9	10	.33
LI2BL	LPC vs PT	1.17	2	6	.14
LCMD	LPC vs PT	1.78	2	3	.11
	LPC vs SL	1.24	14	2	.03
	PT vs SL	2.21	14	3	.15
LCBL	LPC vs PT	2.38	1	1	0
	LPC vs SL	1.94	5	1	0
	PT vs SL	4.63	5	1	0
LP1MD	LPC vs PT	1.21	6	12	.32
	LPC vs SL	1.36	38	6	.27
	PT vs SL	1.66	6	12	.45
LP1BL	LPC vs PT	1.44	7	9	.34
	LPC vs SL	1.59	35	7	.37
	PT vs SL	2.28	35	9	.24
LM1MD	LPC vs PT	1.28	12	3	.09
	LPC vs SL	2.69	36	3	.19
	PT vs SL	2.10	36	12	.72
LM1BL	LPC vs PT	1.09	13	3	.08

Note: Ratio = variance ratio for groups being compared, num d.f. = numerator degrees of freedom, den d.f. = denominator degrees of freedom, CI 95%- = lower bound of 95% confidence interval

[a] LPC = Aggregate Late Precontact, PT = Patale, SL = San Luis.

mechanism; and (3) population size declined and population aggregation increased, thus canceling one another.

Regarding the first proposition the transition from the precontact to the historic period may be difficult to model with the available samples. For most tests, I chose to combine the small precontact samples (Killearn Borrow Pit, Snow Beach, Waddell's Mill Pond) with the Lake Jackson mound to form an Aggregate Late Precontact sample that is diverse both temporally (spanning several hundred years) and geographically (in comparison to the restricted

Table 6.7. Van Valen's Test ANOVA for Apalachee Province samples

n=24	Multiple R = .606		Squared Multiple R = .368		
Source	Sum of squares	df	Mean square	F	p
Sample	1.717	2	.859	6.10	.008
Error	2.953	21	.141		

Matrix of pairwise mean differences

	Lake Jackson	Patale	San Luis
Lake Jackson	0.0		
Patale	–0.91	0.0	
San Luis	0.56	0.58	0.0

Matrix of pairwise probabilities (LSD/Bonferroni)

	Lake Jackson	Patale	San Luis
Lake Jackson	1.0		
Patale	.912/1.0	1.0	
San Luis	.007/.022	.006/.017	1.0

burial catchment of an Apalachee mission cemetery). Patale represents only 17 years of burial activity (Jones et al. 1991) for a population living at the doctrina or within its jurisdiction. Therefore, by nature of the sampling design, I may have artificially inflated, not underestimated, the variance for the precontact populations, thus making a temporal decrease in genetic variability a robust result. Therefore the first interpretation seems unlikely.

Assuming the Patale and precontact samples are drawn from roughly comparable populations, the results indicate little change in genetic variability in the early mission period. There is no evidence for reduction in variability as expected under a genetic drift model or an increase in variability as expected under a model of population aggregation and genetic admixture with formerly isolated populations. For Apalachee, then, incipient missionization effected little change.

There is little comfort in the third interpretation listed because it requires an unlikely convergence of evolutionary mechanisms. In addition, there is little in the historic record to suggest that in-migration (Type 1 and 2) was a significant feature of the early mission period in Apalachee. This result, if upheld from a sampling perspective, most likely suggests reconsideration of the demographic data (proposition 2).

Hann (1988:160–180) provides the most current discussion of temporal population size estimates for Apalachee with consensus estimates as follows: a precontact population size of 25,000–30,000 individuals declines to approximately 16,000 by 1633 (the beginning of organized missionary efforts in Apalachee), which declines to approximately 7,000 individuals by 1650 (the approximate year in which Patale was relocated). Given this degree of depopulation, the precontact aggregate sample should manifest genetic variability consistent with a population twice the size of the Patale congregation. While the data presented here cannot evaluate the size of the population at any particular point in time, it can inform about the rate of change in population size between successive time periods. And, assuming that the sampling design is valid and that phenotypic variance remains proportional to genotypic variance, the results indicate a more limited change in population size. The most parsimonious explanation is that the assumed ubiquitous and rapid decline in population size throughout the New World does not occur in Apalachee, at least initially, or at least at mission Patale.

Reasoned consideration of the evidence suggests that this may in fact be the case. Hann assures us that demographic collapse was a feature of 17th-century Apalachee Province, providing the following reasons: "devastation . . . is known to have occurred from periodic epidemics, the stresses imposed by the Spaniards' labor demands, and the social turmoil resulting from the disruption of the Indians' traditional pattern of living" (Hann 1988:163); however, this statement may relate specifically to post-1650 populations when social changes are known to have commenced in response to demographic collapse in the eastern provinces. In addition, much of the comparative data for demographic collapse assumes models derived from contexts in South and Central America. There are several reasons why these models may be inappropriate: (1) contact with the Apalachee was less prolonged and exploitative; (2) Apalachee was less densely settled than more urban environments in other parts of the Americas (e.g., Mexico); (3) early contact between the Apalachee and Europeans was of a sporadic and violent nature precluding intimate contact necessary for pathogen transfer (notwithstanding the de Soto winter encampment of 1539–40); and (4) disease transmission may have been hampered by unpopulated geographic buffer zones isolating rival 16th-century chiefdoms (Hann 1988:163, 181; DePratter 1991). That these principles do not also apply to the disease-ravaged Guale and eastern Timucua remains problematic.

In Hann's discussion of the population size data, one may detect a certain reticence about the subject matter. And, as I have done previously (Stojanowski 2001), Hann (1988) may assume a priori that epidemic disease was a feature of the early contact period experience, but he does not do so in previous work:

> The 1657 visitation record and the other pertinent documents from the era give no indication that Apalachee was experiencing the calamitous demographic dislocation and decline that was manifest in Timucua. Whereas this trend would continue for Timucua, as the number of its missions shrank with each successive listing . . . all the Apalachee missions existing in 1657 were to survive till the eve of the province's destruction in 1704. (Hann 1986b:392)

Jones et al. (1991) and Storey (1986) likewise assume early and ubiquitous demographic collapse. However, considerations of those statements that promote such a position seem notably unsubstantiated (discussed further below). To the contrary, there are many independent data sources that corroborate the paleogenetic data.

Lack of evidence for pre-1650 Apalachee epidemics. There are no epidemics recorded in the pre-1650 time period for Apalachee. Although the quality of the original sources varies considerably,[1] historians recognize a number of epidemics during the contact period: 1526 (Elvas 1907:173); 1569–70 (Barcía 1951:153; Zubillaga 1946:416); 1582, 1613–17 (Bushnell 1982:13; Deagan 1978:94; Hann 1996:174; Swanton 1922:337); 1649 (Deagan 1978:94); 1650 (Deagan 1978:94); 1653 (Dobyns 1991); 1655–57 (Hann 1988:175); 1659 (Bushnell 1982:13; Dobyns 1983, 1991; Swanton 1922:337); 1670 (Bushnell 1982:13; Swanton 1922:337); 1672–74 (Bushnell 1982:13; Deagan 1978:94; Swanton 1922:337); 1675 (Dobyns 1983:282); 1693 (Hann 1988: 175); 1703 (Hann 1988:175). Although Hann (1988) feels that early epidemics among the Timucua must have affected the Apalachee, he also questions the suppositions of Dobyns (1983) regarding the timing of these epidemics. In fact, Hann explicitly claims that Dobyns's reconstructions are weak at best. Only three epidemics—in the years 1655, 1693, and 1703—are known for Apalachee (Hann 1988: 175), and they all postdate the Patale mission. Hann comments, "there is surprisingly little documentary evidence about the epidemiological mechanisms by which Apalachee's and Florida's populations were reduced so drastically" (Hann 1988:175). Indeed, perhaps this lack of evidence reflects historical reality.

Patale mortuary structure. Consideration of the mortuary structure of the completely excavated Patale cemetery does not present a picture of a population experiencing high mortality. Graves were arranged in nine rows separated by a center aisle, and, with the exception of row nine, none of the rows was completely filled at the time of abandonment. There were equal numbers of left and right side interments, all demographic (age/sex) cohorts were represented, subadults were frequently buried next to adult females, and there were only four graves (numbers 29, 30, 55, 56) that intruded into existing graves (Jones et al. 1991). The implication is of a carefully planned and executed mortuary program not overburdened by excessive mortality. Although I recognize problems with associating mortuary patterns with epidemiological features of past societies (see Henige 1998) (e.g., mass grave = epidemic death), the order and structure of graves at Patale is not what is expected of a society experiencing rapid decline and social upheaval.

Patale mortality rate. Jones et al. (1991) report the complete excavation of the Patale cemetery and identified 67 individuals. Accepting 1633 as the known start date for burial accumulation and approximately 1650 as the date for the abandonment of the Patale cemetery (at 8Le152) based on majolica (Spanish ceramic) inventories (Jones et al. 1991) results in an average of 3.94 burials per year. If the 1675 Florencia enumeration is correct, 500 persons[2] fell under Patale's jurisdiction (Jones et al. 1991:12), and the burial rate (0.8 percent) is not what one would expect of a virgin-soil population ravaged by newly introduced European pathogens. Compared to the 900 interments (210 excavated) at mission San Luis accumulated between 1656 and 1704 (Larsen and Tung 2002), the yearly burial rate at Patale is one-fourth that of the late mission period San Luis cemetery (18.75/year, for 48 years). The San Luis cemetery was also extremely overcrowded, and commingled and disturbed interments were commonplace (Larsen and Tung 2002). In order for burial rates at Patale and San Luis to have been equal, Patale could have been used for only 3.5 years before abandonment.

Patale paleopathology. In light of the previous discussion, the summary of pathological conditions observed in the Patale cemetery (Storey 1986; Jones et al. 1991) are striking. Although raw data are not presented, Jones et al. comment, "the population appears to have consisted of healthy, relatively well-nourished individuals. There is little evidence of infection or anemia. Few teeth have hypoplastic defects . . . but dental caries are common" (Jones et al. 1991: 115). Storey (1986) compared pathological frequency data for Patale and Lake Jackson and concluded similarly, "hypoplasias of the teeth and marks of infection and anemia reveal that the Patale skeletons enjoyed generally *better* health than even high status Pre-Columbian individuals" (Storey, 1986:268, emphasis added). In keeping with established positions on the effects of epidemics, epidemiological interpretations are offered as an explanation for the apparent discrepancy. Interestingly, Jones and colleagues offer a preface to the yet-unpublished osteological paradox of Wood et al. (1992): "otherwise fit individuals were dying from introduced European diseases to which they had no immunity" (Jones et al. 1991:115), and Storey similarly emphasized a shift from chronic to acute stressors during the contact period. While hidden heterogeneity of risks is an acceptable interpretation, at face value these pathological data bolster the argument presented here that epidemic effects may have been overestimated. Concern with the interaction of dental attrition and opportunities to observe hypoplastic defects, and the poor skeletal preservation (see Jones et al. 1991: Figure 55), temper this conclusion. Nonetheless, temporal comparisons with Lake Jackson should be valid, given the similar skeletal preservation and dental wear profiles for both samples (Storey 1986).

If population sizes did remain relatively static during the 16th and early 17th centuries, then either the accepted precontact estimate (25,000 in 1550) is overstated or the early mission period estimate (16,000 in 1633) is underestimated. Although I cannot definitely state whether precontact population size estimates for Apalachee were closer to 30,000 or 15,000 individuals, the phenotypic variance results suggest that whichever is more accurate is likely to apply from the precontact period through 1650.

I hesitate to comment on data in the paleographer's realm but feel some reevaluation is due. Precontact estimates of 25,000–30,000 individuals derive from two sources: Martín Prieto's 1608 estimate of 30,000 individuals is based on an assembled crowd size (Hann 1988:162; Oré 1936:116), and a 1617 estimate is attributed to Fray Luis Gerónimo and is based on an unstated accounting method (see Hann 1988:164). These sources have become widely accepted (Hann 1988; Milanich and Fairbanks 1980). To the contrary, estimates from the 1630s postulate 16,000 individuals based on Governor Horruytiner's enumeration and a 1638 church census referenced in a 1676 document (Hann 1988:164).[3] Weighing the veracity of the pre-1630s and 1630s population size figures is difficult; both are subject to criticism. It seems probable, however, that the 1630s estimates are more in line with the post-1650 epidemic record and better substantiated population size estimates at the close of the 17th century.[4] Because only three epidemics are recorded and Apalachee never experienced significant Stage 1 or 2 congregación (local village and doctrina consolidation) or Stage 1 reducción, the mortality rate in Apalachee was certainly lower than in contemporaneous populations in Guale and Timucua. This discussion does not necessarily falsify demographic collapse. Rather it confirms suspicions that demographic collapse was a more localized phenomenon that affected different areas at different rates and at different times.

Late Transitions

Analysis of phenotypic data bridging the early (Patale) and late (San Luis) mission periods indicated an increase in genetic variability in the latter population. This difference is not compatible with a genetic drift model associated with demographic collapse and is most compatible with evidence suggesting population aggregation or admixture increased in frequency in Apalachee Province during the later 17th century.

Modeling the transition to the late mission period is in some ways easier and in some ways more difficult than for the precontact-Patale comparisons. Patale and San Luis were both mission doctrinas with similar functions in terms of mortuary practices within their respective communities. This obviates concerns with differential site function (small lineage mound vs. large community cemetery) that applied to discussions of the precontact-Patale transition. However, San Luis was a different type of mission center than Patale by virtue of its status as the provincial capital housing a large nonindigenous population (Boyd et al. 1951; Hann 1988; McEwan 1991a, b, 1992, 1993, 2000, 2001). On the other hand, Patale was a distinctively Native American settlement with only a single resident friar, a lack of military presence, and no Spanish secular officials or citizenry (Jones et al. 1991). Because I cannot parcel the effects of time and function when comparing San Luis to Patale, it is difficult to discern what exactly is initiating the transformation in variance patterns after 1650. Causes may be either functional (representing different community structures at San Luis and Patale) or temporal (reflecting changes typical of the late mission period Apalachee, in general). That San Luis experienced greater aggregation or admixture with nonlocal populations is consistent with the research models developed in this book (chapter 3). Discerning whether these changes are applicable to *all* late mission period Apalachee doctrinas or just to San Luis remains problematic.

More has been written about San Luis than any other Apalachee mission, owing in part to the long-term interdisciplinary research project developed to investigate the social history of this frontier settlement (Boyd et al. 1951; Cordell 2002; Hann 1988; Hann and McEwan 1998; Larsen and Tung 2002; Larsen et al. 1996; McEwan 1991a, b, 1992, 1993, 2000, 2001; Reitz 1993; Ruhl 2000; Scarry 1993; Shepard 2003). Because it was the provincial capital housing secular Spanish officials, multiple friars, and a Spanish garrison, the historical records are particularly rich for this site (Boyd et al. 1951; Hann 1988; Hann and McEwan 1998; McEwan 1991a, b, 1992, 1993, 2000, 2001).

Although San Luis may have differed from surrounding settlements in numerous ways, the most obvious and significant difference was the presence of Spanish citizens and soldiers that resulted in the development of a biethnic, integrated community by the time of the mission's abandonment in 1704. Hann (1988:200) summarizes the historical presence of Spanish soldiers at San Luis and indicates a garrison population that varied between 12 and 45 soldiers during the latter half of the 17th century. Hann and McEwan (1998:63) suggest that by the end of the 17th century several hundred Spanish soldiers and citizens were living at or near San Luis. It is known that some of the soldiers found wives among the native Apalachee (Boyd et al. 1951:24–26), which resulted from both the short supply of unmarried Spanish females in the colony (McEwan 2001:34) and the active acceptance and promotion of mestizaje as a means of effecting the acceptance of a Christian lifestyle (McEwan 2001). Archaeological analyses of material culture distributions and settlement design additionally suggest that a burgeoning mestizo or Hispanic population was in residence (McEwan, 1991a, b, 1992, 1993, 2000, 2001). Hann (1988) cites numerous circumstantial examples suggesting that Spanish-Apalachee mestizaje was commonplace: (1) the adoption of Spanish names by some ethnic Apalachee (Hann 1988:252), despite the general trend of indigenous name maintenance by Apalachee elite (Hann 1988:251);[5] (2) Franciscan requests for the removal of Spanish soldiers to alleviate concerns of Apalachee males about their wives and daughters (Hann 1988:252); (3) post-1700 emigré statements about concubinage that had occurred while they lived under Spanish authority (Hann 1988:170); (4) governmental proclamations condemning concubinage, suggesting that it was a significant problem and common behavior (Hann 1988:170).

It is worth noting that Spaniards preferred to be buried in St. Augustine (McEwan 2001:641), and ongoing paleogenetic analyses of grave structure at San Luis have failed to identify Spanish burials definitively (Stojanowski et al. n.d.). Despite documentation of pathology and isotopic patterns more typical of non-Apalachee (Larsen et al. 1996), that is, the rarity of caries and limited isotopic data suggesting that maize was not consumed by some elites on a daily basis (Larsen et al. 2001), there is no evidence that the Spanish soldiers or citizenry would have been buried in the San Luis church. Such a practice would have artificially inflated variance estimates.

In addition to the potential effects of the Spanish presence, the lives of San Luis's converts were also affected by two processes common to all post-1650 Apalachee communities: changes in New World political landscapes that accelerate after 1650, in particular a stronger English presence and resulting ramifications, and demographic collapse in communities around St. Augustine and among coastal Guale and St. Johns River Timucua chiefdoms that was nearly complete by this time period (Worth 1995, 1998a, b).

The English bid for North American dominance, beginning in earnest with the establishment of Charlestown in 1670, initiated far-reaching changes in Native American social and political structures throughout eastern North America. Freshly provisioned with arms and positioned strategically between both two- and three-way power struggles over North America (English-French-Spanish), Native American populations became involved in slave raiding and other activities deleterious to both the offender and the offended communities. This social upheaval sent shock waves throughout Southeastern chiefdoms already stressed by internal pressures brought on by epidemics. By choice, many populations sought protection under the Spanish flag. Apalachee, one post-1650 nexus of interaction, received many of these immigrants.

Hann summarizes known examples of nonlocal immigration into Apalachee that included both Type 1 (Christianized) and Type 2 (non-Christianized) reducción populations. As discussed in chapter 3, the political position of Apalachee during the time of the earliest 16th-century entradas strongly suggests biological divergence between Apalachee and its neighbors, thereby making the in-migration of these populations a biological process with expected variance increasing repercussions. Of the former type, western Timucua populations (Utina and Yustaga), displaced after the disastrous 1656 interior uprising and the reprisals that followed, are likely candidates (Hann 1996; Worth 1998a, b). The problem was severe enough in 1657 to elicit Governor Rebolledo's intervention:

> I order and command all the Indian men and Indian women of the Provinces of Timucua and Ostaca who at present are to be found in this [province] of Apalache to leave it and retire to the places of which they are natives to be under the obedience of their caciques within the fifteen days that they are given as a limit. This order does not embrace those who are domiciled [here] two years. I condemn the men who do not comply with what has been ordered to one hundred lashes and four years forced labor for each one, and the women to one hundred lashes and that they serve in the *presidio* of St. Augustine. (Hann 1986a:102, 103)

There is no mention of Guale in-migration that I am aware of; however, Swanton mentions emigration of Guale to as far south as the Calusa territory on the south Gulf Coast (Swanton 1922:343). Of the latter type of immigrants (Type 2), Hann (1988) records numerous examples: 300 Tocobaga from as far south as Tampa Bay (Hann 1988:165); ethnic Chine, Yamassee, and Chacato living within San Luis's jurisdiction (Hann 1988:102, 103); an Oconee mission established during the 1650s (Hann 1988:179); and Tama and Pacara, similarly recorded in a 1674 census (Hann 1988:322). In-migration was so extensive during the late 17th century that census sizes by mission actually increase for some doctrinas, despite documented epidemics in the province (Hann 1988:173–174; Milanich 1996). In 1675, 10 percent of the population was non-Apalachee (Hann 1988).

Repercussions of population displacement were paralleled in severity by changes internal to the Spanish system. Demographic collapse in the eastern provinces necessitated reallocation of labor demands, and the comparatively populous Apalachee region received greater attention from the colonial government as a result (Hann 1986b, 1988). After 1650, Apalachee involvement in the sociopolitical aspects of the colony increased as Spanish officials and settlers redirected efforts to create a productive economy farther west. Interior ranches appeared, trade with Havana via the St. Marks River increased, and repartimiento obligations for Apalachee males became onerous. All three mechanisms initiated the gradual removal of the Apalachee male from the normal course of social involvement. Male transiency, therefore, became problematic as the burgeoning cash economy ranching industry, the required servitude in St. Augustine, and the avoidance of required work quotas altogether created a male demographic vacuum in the province (Hann 1988:146–147). At least two sources indicate that this was a serious and recognized problem (Bushnell 1979:5; Hann 1988:171). With males absent and traditional mating networks approaching inoperability, it is reasonable to assume that females would take advantage of marriage opportunities with both Spanish military and resident, nonlocal indigenous groups.

Therefore, the results of this analysis are completely congruent with ethnohistoric predictions. Mestizaje with Spanish soldiers and interbreeding with ethnically distinct Native American populations are both likely sources of increased variability. However, emphasis on amplifying mechanisms belies the contradiction that population sizes are known to have declined between 1650 and 1704. Why then do we not see evidence for a decline in phenotypic variability in the bioarchaeological record? This final point is worth additional consideration.

Hann's (1988) summary of the population size data for Apalachee suggests a definitive decline in population size after 1650. Accepting the 1633 estimate of 16,000 as a baseline, the most reliable estimates suggest a decline to 10,500 by 1675 and a further decline to around 8,000 individuals by the time the province was abandoned (Hann 1988:162). In addition, direct evidence for epidemics, so lacking for Patale's converts and contemporaries, has been documented for Apalachee in the years 1657, 1693, and 1703, the last referring to an epidemic at San Luis itself (Hann 1988:175). Because the site sampled in this analysis (San Luis de Talimali) is believed to have been established in the year 1656 (McEwan 1991a, b, 1992, 1993, 2000, 2001), the effects of these epidemics should have been manifest in the San Luis population. Why, then, is there no evidence for population size decline in the burial record, and, further, does this suggest additional reconsideration of the demographic data?

Besides the historical estimates of changes in population size, the best data we have that suggest that population sizes were declining come from two sources: mortuary patterning indicating an increase in crowding and commingling, which are suggestive of an increase in the rate of death, and bioarchaeological data examining changes in population stress and quality of life.

As discussed, the mortuary structure at San Luis is consistent with a population experiencing high mortality. Although I know of no mass graves, secondary and commingled interments were commonplace, and the rate of burial approached 19 per year, on average (Larsen and Tung 2002). Accepting a population size of 1,400 individuals that lived within San Luis's jurisdiction (McEwan 1991a:38), the rate of burial is approximately 1.4 percent per year, which is roughly similar to the adjusted 0.8 percent per year for Patale. Although these very crude figures are based on even cruder population size estimates and assumptions about the period of interment activity and the complete recovery of burials (see Larsen and Tung 2002:24), burial density may belie the simple position that crowded and commingled cemeteries such as that documented for San Luis, Santa Catalina de Guale (Larsen 1993), and the presumed secondary or tertiary location of the Patale congregation at the O'Connell mission site (Marrinan et al. 2000) represent an increase in mortality rates. In fact, the size of the population served by the cemetery and the duration of use may be primary to epidemiological considerations. Certainly it seems that cemetery overuse was not cause for relocation of a doctrina.

In addition, bioarchaeological data for Apalachee, though less comprehensive than for the Guale, does not indicate similar levels of systemic physiological stress. As discussed above, with the exception of a high caries prevalence, Patale individuals demonstrated less evidence for metabolic disturbance than even elite precontact Apalachee (Jones et al. 1991). Similar data can be marshaled for San Luis.

(1) Caries rates, typically associated with a high maize diet and with declining health conditions, were low at San Luis, only 4.5 percent (Larsen and Tung 2002), which is much lower than comparative data from other mission contexts (Larsen et al. 1991).

(2) Collagen preservation was poor, prohibiting the extraction of stable isotope data from all but one individual. This individual, a high status individual buried near the altar end of the church, demonstrated isotopic signatures *inconsistent* with the consumption of maize on a daily basis (Larsen et al. 2001).

(3) Data on the frequency of pathological striae of Retzius, which target early infant stress events of short duration such as diarrhea bouts, affect the fewest number of individuals at San Luis (67 percent, as measured by percent of individuals with at least one pathological striae) compared to all other mission samples and also demonstrate one of the lowest morbidity rates (2.2) of the fully Christianized comparative mission samples (Simpson 2001).

(4) Analysis of hypoplastic defects in the mission and precontact populations generally supports the above data. In terms of the percentage of individuals affected by hypoplasia (Hutchinson and Larsen 2001:186; Larsen and Tong 2002), San Luis was less affected than all Georgia time periods and less affected than Florida early precontact populations. San Luis was comparable in frequency to Florida late prehistoric samples and demonstrated more hypoplasias in comparison to Patale. For maxillary incisors San Luis exhibited fewer teeth affected in comparison to the Florida prehistoric, Patale, and all Guale samples (Hutchinson and Larsen 2001:187; Larsen and Tong 2002). For mandibular canines San Luis was less affected than all Georgia samples and roughly comparable to preceding Florida samples, with exception, perhaps of Patale. Hutchinson and Larsen conclude that all Florida populations were less stressed than Georgia populations for all time periods.

(5) Osteoarthritis at San Luis was relatively infrequent in comparison to contemporaneous Guale samples. Approximately 40 percent of adults were affected at San Luis compared to over 60 percent affected at Amelia Island (Larsen and Tung 2002:35).

I am therefore left in the same position as with Patale. The evidence for increased morbidity and mortality is simply not there when considered from bioarchaeological and paleogenetic perspectives. And, assuming that population size estimates for Apalachee are correct, interpretations unrelated to epidemiological factors are possible. One of the more likely is documented out-migration. That is, population sizes decline in Apalachee not from excessive disease (although many deaths were surely related to introduced diseases) but from fugitivism, a factor also documented in greater intensity after 1650. As some Apalachee were fleeing because of excessive labor demands levied by the Spanish, other populations, of whom less work was initially expected, begin to enter the province. If the emigrants represented random samples of the population, then no change in population genetic composition is expected.

This discussion aside, assuming that demographic stress was a feature of the San Luis population, there is still good reason *not* to expect evidence of it in the mortuary record. If one considers the relationship between the population that lived at San Luis and the burial sample that resulted from their collective mortuary activities, there is good reason to believe that population size could have declined but not manifested as a decrease in phenotypic variability in the San Luis burial sample. If I am correct that epidemic-related deaths were primarily a feature of the post-1650 Apalachee experience, then accumulation of deaths owing to these mortality events would be coincident with the establishment of the San Luis cemetery included in this analysis. As population size declined throughout the mission period, the population at San Luis would become less variable due to genetic drift. However, everyone who died between 1656 and the 1704 abandonment of the mission would have been buried in the same cemetery. Therefore, the cemetery sample provides little resolution of differences in population genetic variability that developed during the course of the mission's use. Comparison of the initial population (ca. 1656) with that which abandoned the mission (ca. 1704) should demonstrate evidence for demographic collapse, if it occurred. The level of genetic variability in the 1704 San Luis population would be represented in the succeeding cemetery, which does not exist. Having said this, there is no guarantee that the loss of variation due to drift would counteract the increase in variation from admixture and population aggregation. In other words, the analytical design developed in this book has greater sensitivity to detect increases in genetic variability but more limited sensitivity to detect decreases in genetic variability, particularly if the decline occurred during the period in which the cemetery sample was in the process of active formation.

The complex interpretations of transformations in population genetic variability in three temporal periods in Apalachee Province share one theme. Assumed demographic collapse may not be completely justified for Apalachee. The position most well supported by this research is that the Apalachee were genetically distinct from their neighbors during the 16th century, that after 1650 many immigrant groups settle in the province, and that admixture likely occurred between these populations, resulting in a diverse multi-ethnic community at the time of the English-sponsored raids on the province.

Despite the abundance of archaeological and historical information available for the Apalachee, the bioarchaeological resources from this province are scarce. While both the Patale and San Luis cemeteries are of moderate and representative size and are moderately preserved (dentally), there are no similarly large precontact samples from the survey region. Therefore, an aggregate approach was adopted. Data from four precontact samples were used to generate variance estimates for an aggregate spatially and temporally variable population. Such aggregation of disparate samples is expected to lead to a bias toward greater population genetic variability than was actually present in the late precontact Apalachee.

Despite this potential for bias, the statistical analyses indicated no evidence for an increase or, more importantly, a decrease in phenotypic variability associated with the transition from the late precontact to mission periods. This result is particularly surprising, given the composition of the precontact sample, and may suggest that the effects of drift were more limited than hypothesized. In particular, the case for Apalachee epidemics before 1650 may require reconsideration.

The transition from the early to late mission period indicated that phenotypic variability was at a maximum at the post-1650 San Luis mission despite the fact that population size in Apalachee was lowest during this time period. Although this result may represent the effects of processes affecting the entire post-1650 Apalachee populace, this is impossible to determine without examining additional post-1650 Apalachee mission samples. Increased variability at San Luis more likely results from the unique community residing at this mission, which included Spaniards and native Apalachee, as well as immigrant indigenous populations. The last reflects population movements after the mid-century turmoil in the western Timucua districts, and the southward immigration of interior groups in the wake of English-sponsored slave raiding. Despite estimates of precontact population size and the 73 percent nadir proposed for Apalachee, there was no evidence of the expected effects of genetic drift. Stasis and genetic admixture, therefore, assume a primary position in the population history of the Apalachee.

Notes

1. Hann is particularly critical of Dobyns's use of Florencia's rounded population size estimates as evidence of an epidemic in progress. In Florencia's words, "I have not taken a census and they die daily" (Florencia in Hann 1988:162).

2. Although Florencia's estimate refers to the Patale congregation in the 1670s, thereby a different physical mission than the 8Le152 sample, it seems reasonable to assume, given statements about Apalachee demography, that the pre-1650 Patale jurisdiction was that large or larger. If larger, the role of burial was even lower.

3. Swanton (1922:120) indicates that the latter derives from a memorial for Fray Alonso del Moral dating to November 5, 1676, in which it was stated that the population of Apalachee was 16,000 in 1638 and that at the time of writing, 1676, there were only 5,000 remaining.

4. It is interesting to note the change in sentiment regarding the size of the Apalachee population over the years. In 1922 Swanton cited a November 1633 letter indicating there were 15,000 to 16,000 Apalachee, which Swanton considered unrealistically high but less so than the claims of 30,000 made in 1618 and 34,000 made in 1635 (Swanton 1922:18). He was of the opinion that fewer than 5,000 individuals were present in the province. Lanning (1935:166) held a similar opinion.

5. Hann (1988:286) noted that it was unusual that most Apalachee had maintained their indigenous surnames when congregated near St. Augustine after the missions were destroyed. Other ethnic groups had adopted Spanish surnames.

7 Aggregation and Collapse on the Georgia Coast

To generate predictions for the Guale, I must consider the previous work of Griffin and colleagues (Griffin 1993; Griffin et al. 2001) that documented an initial increase in trait variability during the early mission period followed by a dramatic decline in trait variability during the later years of the Spanish period. These data were interpreted as reflecting the processes of population aggregation at Santa Catalina de Guale during initial phases of missionization, followed by a dramatic decline in variability coincident with a population bottleneck that occurred sometime during the 17th century. The demographic data for Guale summarized in chapter 4 do suggest a rapid and catastrophic loss of population numbers with an attrition rate of nearly 99 percent. Griffin's results, therefore, are consistent with the ethnohistoric record.

Based on ethnohistoric discussions of the Guale, their relationship with neighboring groups, and their position in regional political affairs, I can also generate expectations based on the model developed in chapter 3. As with the Apalachee, both Stage 1 (consolidation to the doctrina) and Stage 2 (doctrina consolidation) congregación would have had limited effect on variability levels *within* Guale province because of the presumed cultural homogeneity along the Georgia coast. However, unlike among the Apalachee, these aggregation efforts were used extensively because population size declined rapidly after initial contact in the early 16th century. However, the hypothesized effects of this form of population aggregation are minimal for the same reasons as outlined for the Apalachee: cultural and linguistic homogeneity within the borders of the Guale territory. As with Apalachee, linguistic and political differentiation suggests biological circumscription of Guale; they warred with their northern Orista neighbors and were at war with and linguistically isolated from their Timucua neighbors to the south. This suggests that Type 1 reducción (Christian in-migration, e.g., the conflation of Mocama and Guale) and Type 2 reducción (pagan in-migration, e.g., the sudden appearance of the Yamassee after 1650) *would* have significantly increased variability levels.

Differences between the Guale and Apalachee mission experiences relate to the timing of the demographic transition. Ethnohistoric and bioarchaeological data suggest that interethnic contact and aggressive population aggregation occurred much earlier in the coastal provinces because of the immediate effects of epidemic disease in local populations brought on by early slave raiding activity along the Atlantic coast. The Spanish presence at the Santa Catalina de Guale mission in the early mission period may have provided, as with San Luis, another vector for increased phenotypic variability. However, the process of depopulation was a continuous one, and after a certain point aggregation ceases as small remnant populations continue to experience demographic stress. I predict, therefore, that variability increased with the transition from the late precontact to early mission period (1600–1680) owing to aggregation and the potential of Spanish admixture, a process similar to that documented for San Luis. The late mission period, however, should demonstrate a decline in phenotypic variability because of the effects of genetic drift in populations rapidly decreasing in size.

Bioarchaeological Resources

The Guale region is the best represented province in terms of the quality of skeletal material (Table 7.1, Figure 7.1). There are two mission period samples, Santa Catalina de Guale (SCDG) and Santa Catalina de Guale de Amelia (Amelia), that are affiliated in an ancestor-descendant relationship. In other words, the SCDG sample is the biological and historical predecessor of the Amelia Island sample, which obviates concerns with lineage consistency that existed for Apalachee. These samples, therefore, represent a continuous biological lineage that allows definitive estimation of changes in population genetic variability *within* the same population (Larsen 2001).

Santa Catalina de Guale is an early mission period site located on St. Catherine's Island, Georgia. Non-cemetery excavations were conducted by David Hurst Thomas of the American Museum of Natural History for over a decade beginning in 1974 (Thomas 1990b). Cemetery excavations were conducted by Clark Spencer Larsen between 1982 and 1986 (Larsen 1990, 1993). Burials derive from a church structure believed to date to the 17th century, likely representing the church rebuilt in 1607 after the first was destroyed during the Guale uprising a decade earlier (Thomas 1990b). Therefore, burial activities commenced in 1607 and ended in 1680 (Larsen 1990), when the site was abandoned and its population aggregated with another doctrina farther south (Worth 1995).

Most burials were single, extended, and supine interments with heads oriented east-southeast. Hands were often clasped over the chest in a folded po-

Table 7.1. Guale Province skeletal samples

Sample	Size	Period	Source
Santa Catalina de Guale	432	Early Mission	Larsen 1990
Santa Catalina de Guale de Amelia	121	Late Mission	Larsen 1993
Irene Mound and Mortuary	248	Late Precontact	Caldwell and McCann 1941
Kent Mound	25	Late Precontact	Larsen 1982
Pine Harbor	50	Late Precontact	Larsen 1990; Cook 1980
South End Mound	15	Late Precontact	Larsen and Thomas 1986; Larsen 1982
Johns Mound	60	Late Precontact	Larsen and Thomas 1986; Larsen 1982
Marys Mound	10	Late Precontact	Larsen and Thomas 1986; Larsen 1982
Seven Mile Bend	6	Late Precontact	Larsen 1982
Lewis Creek	9	Late Precontact	Larsen 1982
Little Pine Island	7	Late Precontact	Larsen 2001

sition. Burial features were aligned parallel to the long axis of the church in orderly rows, which suggests non-epidemic interment (Larsen 1990, 1993). The prevalence of disturbed burials does suggest, however, a high mortality experience for this population. Larsen (1990) documented no patterning of burials by age or sex, but grave goods were disproportionately located in graves near the altar end of the church (Larsen 1990), suggesting status variation in the population.

A total of 432 individuals were recovered, although most were too fragmentary to be included in this analysis. The Santa Catalina sample represents one of the few completely excavated cemeteries from this time period such that the reported MNI is an accurate estimate of the total number of burials within the church walls (Larsen 1993). Because all nondental remains were repatriated prior to my analyzing the collection, age and sex estimates were based on observations by Larsen and colleagues (Larsen 1990). Age estimates are, however, very reliable (Larsen personal communication, 2000) and are based on complete seriations of the dentition and multifactorial aging techniques (Russell et al. 1990).

Santa Catalina de Guale de Amelia represents the relocated mission population of the St. Catherine's Island (SCDG) doctrina, which moved south to Amelia or Santa María Island (Worth 1995). While the site has been designated

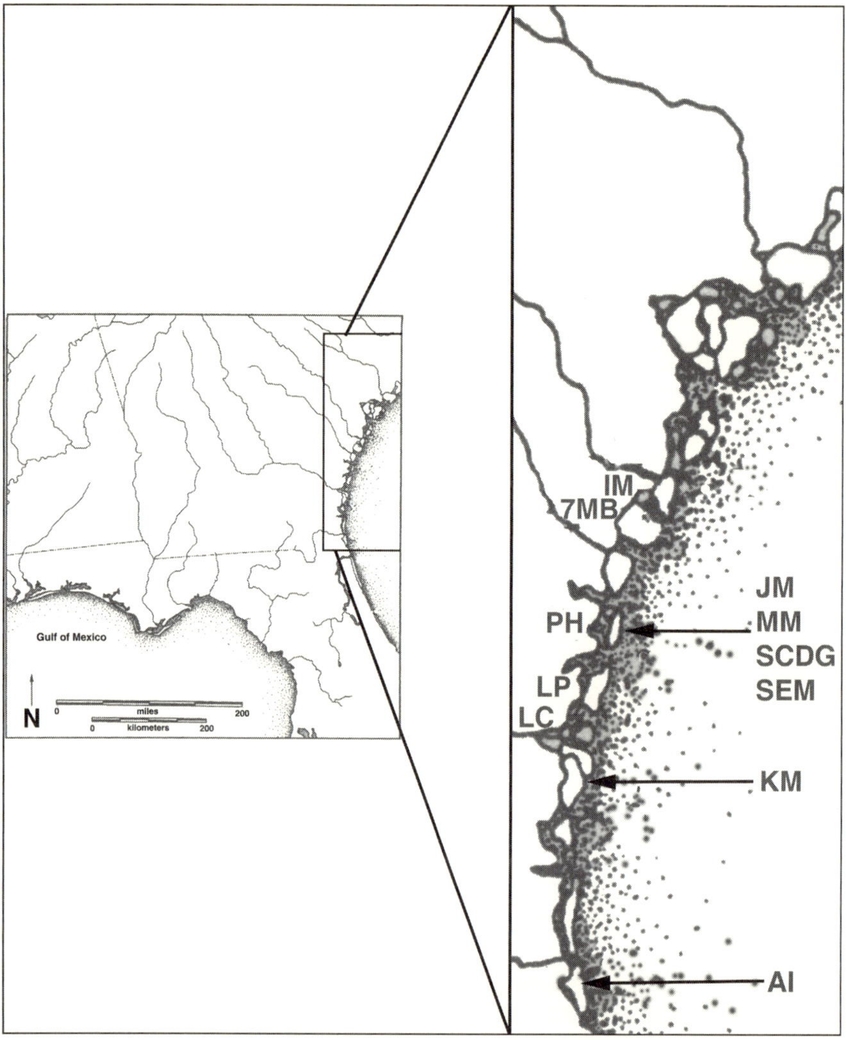

Figure 7.1. Map of Florida showing locations of the Guale samples used in this analysis. IM = Irene Mound and Mortuary, 7MB = Seven Mile Bend, PH = Pine Harbor, LP = Little Pine Island, LC = Lewis Creek, JM = Johns Mound, MM = Marys Mound, SCDG = Santa Catalina de Guale, SEM = South End Mound, KM = Kent Mound, AI = Santa Catalina de Guale de Amelia Island (modified from Larsen 2001: Fig. 2.1).

multiple names by different excavators (see Larsen 1993:326), it was positively identified as mission Santa Catalina when the seal of the mission was located during field excavations (Hardin 1986). The temporal position of the skeletal sample is, therefore, well established and encompasses a brief period of interment activity (between 1686 and 1702) (Larsen 1993).

Individuals were interred within the walls of a relatively simple church structure. A total of 121 burials were recovered, most well preserved (Larsen 1993). As with other mission churches, the burials were aligned parallel to the long axis of the church and were interred in an extended and supine position with hands clasped across the chest. Most burials (93 percent) were undisturbed single interments, suggesting a short period of use not associated with high mortality rates (Larsen 1993). There were few grave goods found in association with the burials (Larsen 1993).

Bioarchaeological materials from the precontact period were also plentiful, with multiple samples available from the same geographic area as the mission period samples. Irene Mound and Irene Mortuary represent an earlier and later component of a single large aggregate ceremonial complex located inland and just north of traditional Guale territory. In addition, there are multiple samples that derive from the coastal islands or just inland of areas associated with the Guale (Kent Mound, Pine Harbor, South End Mound, Johns Mound, Marys Mound, and Seven Mile Bend). Because the samples are roughly contemporaneous and derive from a limited geographic region, they are pooled into an aggregate late precontact (LPC) sample for the purposes of comparing variability differences.

The *Irene Mound* site is a late prehistoric multiple mound complex located on the Savannah River in Chatham County, Georgia. Excavations by Joseph Caldwell between 1937 and 1939 produced the skeletal material included in this study (Caldwell and McCann 1941). Human remains were recovered from two locations, a small burial mound and a large mortuary complex physically distinct from the mound (Caldwell and McCann 1941). Construction of the mound commenced around A.D. 1200, and burial activity continued in the mound for perhaps 200–300 years. Many of these burials were primary, flexed, single interments though considerable variability in burial position was recorded (Caldwell and McCann 1941). The mortuary structure is a later and physically separate cemetery associated with the protohistoric Irene phase (Caldwell and McCann 1941). The mortuary was a rectangular semisubterranean structure made from wattle and daub.

The initial bioarchaeological report on the burial remains was prepared by Frederick Hulse (1941). Hulse recorded 265 individuals, 74 males, 75 females, and the remainder indeterminate. There were 16 recorded adolescents and 38 children and infants. For this analysis the sample sizes were more modest. The mound component has an MNI of 91 individuals; the later mortuary component has an MNI of 34 individuals.

Other prehistoric burial samples were much more modest in size, reflecting a primary difference in usage patterns; smaller mounds likely served a smaller percentage of the overall population. *Kent Mound* is a late prehistoric site on St. Simons Island, Georgia (Larsen 1982:179), dating to the Irene phase

(A.D. 1300–1550). Only eleven individuals were available for observation. Preservation ranged from fair to good, which allowed accurate age and sex estimation. *Pine Harbor* is a late prehistoric to early contact period mound site located 15 kilometers west of St. Catherine's Island on the Georgia mainland (Larsen 1990). According to Cook (1980), Pine Harbor may represent a late precontact to early contact population that experienced a severe epidemic prior to the establishment of missions along the Georgia coast. *South End Mound* is a late prehistoric site located on St. Catherine's Island, Georgia, dating to the Irene phase (A.D. 1300–1550). Only a dozen individuals were available for inclusion in this analysis (Larsen and Thomas 1986). *Johns Mound* is a late prehistoric burial mound located on St. Catherine's Island, Georgia. Radiocarbon dating placed burial activities around A.D. 1119 ±60 years (Larsen and Thomas 1982:297). A total of 69 well-preserved, but fragmentary, burials were recovered from Johns Mound. Most burials were primary interments with a preference for extended, supine burial treatment; however, bundle burials, flexed primary burials, and cremations were also reported (Larsen and Thomas 1982:325). *Marys Mound* is a late prehistoric site located on St. Catherine's Island, Georgia. A single radiocarbon date placed burial activities approximately 700 years ago (A.D. 1255 ±70) (Larsen 1982:176). The skeletal sample consisted of six well-preserved individuals. Fragmentation, however, made age and sex estimation difficult. Although both primary flexed and bundle burials were recovered, there were no cremations or primary extended interments as seen in Johns Mound. *Seven Mile Bend* is a late prehistoric site located near Richmond Hill, Georgia (Larsen 1982:180). Only six individuals were available for analysis, all of which were poorly preserved. The poor condition of the remains made sex and age estimation difficult. The exact temporal affiliation of the burials is uncertain; however, a broad estimate of 1150 to 1550 is suggested by the ceramic assemblage (Larsen 1982:180). *Lewis Creek* is a series of multiple interment mounds located near Darien, Georgia, and dating to the late precontact period (Larsen 1982). No radiocarbon dates were available for this sample, but it is believed to date to the late precontact period. *Little Pine Island* is a small unpublished sample that Larsen (2001) attributes to the late precontact period.

Analytical Results

Univariate Variability Differences

Sample sizes and standard deviations are presented in Table 7.2. Because sample sizes are larger for the Guale database, analyses are repeated for males and females separately in addition to the aggregate sex comparisons. The pattern of standard deviations suggests a pattern of phenotypic transformation that was different from that observed in Apalachee. Comparison of the LPC sample with Santa Catalina de Guale indicated an increase in variability for 7 of 11

Table 7.2. Guale aggregate precontact and mission period sites (sample size, standard deviation)

Measure	LPC	IMd	IMt	SCDG	Amelia
UI1BL	127(.62)	55(.68)	16(.56)	35(.78)	49(.44)
UCMD	116(.54)	57(.51)	17(.39)	98(.44)	44(.56)
UCBL	138(.60)	65(.66)	20(.54)	95(.56)	46(.61)
UP1MD	128(.67)	61(.45)	18(.43)	138(.84)	48(.65)
UP1BL	144(.71)	69(.77)	19(.45)	138(.90)	48(.64)
UM1MD	145(.58)	65(.52)	25(.48)	age effect	47(.84)
UM1BL	184(.63)	91(.58)	32(.62)	144(.73)	51(.61)
LI2MD	88(.48)	45(.45)	15(.61)	81(.91)	—
LI2BL	124(.55)	66(.63)	20(.34)	62(.64)	51(.38)
LCMD	108(.55)	50(.58)	23(.48)	71(.5)	48(.53)
LCBL	138(.67)	66(.61)	29(.73)	62(.60)	55(.65)
LP1MD	131(.51)	68(.49)	age effect	137(1.09)	50(.47)
LP1BL	146(.61)	77(.64)	29(.46)	age effect	52(.43)
LM1MD	160(.55)	73(.50)	31(.70)	age effect	40(.56)
LM1BL	187(.52)	89(.50)	34(.50)	142(.52)	age effect

Note: LPC = Aggregate Late Precontact, IMD = Irene Mound, IMT – Irene Mortuary, SCDG = Santa Catalina de Guale, Amelia = Santa Catalina de Guale de Amelia. "Age effect" indicates the data were excluded because of a significant correlation with age.

measurements, suggesting that early mission period populations were more variable than precontact ancestral populations. Santa Catalina de Guale was also more variable than Irene Mound for 8 of 12 measurements and was more variable than Irene Mortuary for 10 of 11 comparisons. The transition between the early (SCDG) and late (Amelia) mission periods was characterized by a decrease in variability for 6 of 10 measurements.

Results of univariate inferential analyses are presented in Tables 7.3 and 7.4 for the maxillary and mandibular dimensions, respectively. Irene Mound, Mortuary and the coastal samples are considered a single biological population for these tests. Bartlett's tests indicate significant differences in variability for UI1BL, UP1MD, UP1BL, UM1MD, LI2MD, LI2BL, LP1MD, and LP1BL at the 5 percent level and LM1MD at the 10 percent level. Levene's tests indicate significant differences in variability for UI1BL and LP1MD at the 5 percent level and UP1BL at the 10 percent level. P-values for the Bartlett's tests do not correspond with the pattern of nonoverlapping confidence intervals, however, so delineating which comparisons are driving the results was difficult. If I only

Table 7.3. Guale maxillary univariate variance tests

Measure	Period[a]	n	sd	95%CI (sd)	Bartlett Stat (p-value)	Levene Stat (p-value)
UI1BL	LPC	127	.62	.53–.73	13.5(.00)	2.76(.04)
	SCDG	35	.78	.60–1.11		
	AM	49	.44	.35–.59		
UCMD	LPC	116	.54	.46–.64	5.91(.11)	1.40(.71)
	SCDG	98	.44	.38–.54		
	AM	44	.57	.44–.77		
UCBL	LPC	138	.59	.52–.70	.51(.92)	.47(.71)
	SCDG	95	.56	.47–.68		
	AM	46	.61	.48–.82		
UP1MD	LPC	128	.68	.59–.79	7.91(.02)	.64(.53)
	SCDG	138	.84	.73–.98		
	AM	48	.65	.52–.86		
UP1BL	LPC	144	.71	.62–.83	11.44(.00)	2.41(.09)
	SCDG	138	.90	.78–1.06		
	AM	48	.64	.51–.86		
UM1MD	LPC	145	.58	.51–.68	16.60(.00)	2.01(.11)
	AM	47	.84	.66–1.12		
UM1BL	LPC	189	.63	.56–.72	4.42(.22)	.37(.77)
	SCDG	144	.73	.63–.85		
	AM	51	.61	.48–.80		

[a] LPC = Aggregate Late Precontact, SCDG = Santa Catalina de Guale, AM = Santa Catalina de Guale de Amelia.

consider p-values with nonoverlapping intervals, the results support the patterns discussed above. For three measurements the Santa Catalina de Guale sample was significantly more variable than the Santa Catalina de Guale de Amelia sample (UI1BL, LI2BL, LP1MD), and for two measurements the Santa Catalina de Guale sample was significantly more variable than the LPC sample (LI2MD, LP1MD).

Variance ratio confidence intervals are presented in Tables 7.5 and 7.6 for the maxilla and mandible, respectively. None of the maxillary measurements produced significant results. However, the mandibular data produced three significant differences, all of which indicated that the Santa Catalina de Guale

Table 7.4. Guale mandibular univariate variance tests

Measure	Period[a]	n	sd	95%CI (sd)	Bartlett Stat (p-value)	Levene Stat (p-value)
LI2MD	LPC	88	.48	.41–.58	29.14(.00)	.986(.38)
	SCDG	81	.91	.76–1.12		
LI2BL	LPC	124	.55	.47–.65	18.93(.00)	1.03(.38)
	SCDG	62	.64	.52–.83		
	AM	51	.38	.30–.49		
LCMD	LPC	108	.55	.47–.66	2.10(.55)	.79(.50)
	SCDG	71	.57	.47–.72		
	AM	48	.53	.42–.71		
LCBL	LPC	138	.66	.57–.77	3.99(.26)	1.79(.15)
	SCDG	62	.59	.48–.77		
	AM	55	.65	.52–.85		
LP1MD	LPC	131	.51	.44–.59	91.31(.00)	2.81(.04)
	SCDG	137	1.09	.95–1.28		
	AM	50	.47	.38–.62		
LP1BL	LPC	146	.61	.53–.70	16.49(.00)	2.09(.15)
	AM	52	.43	.35–.56		
LM1MD	LPC	160	.55	.48–.64	5.15(.08)	1.72(.18)
	AM	40	.56	.44–.78		
LM1BL	LPC	187	.52	.46–.59	.21(.98)	.09(.96)
	SCDG	142	.52	.45–.61		

[a] LPC = Aggregate Late Precontact, SCDG = Santa Catalina de Guale, AM = Santa Catalina de Guale de Amelia.

sample was more variable than LPC (LI2MD, LP1MD) and Amelia Island (LP1MD).

Sex-Specific Univariate Variability Differences

Univariate variance tests were repeated by sex using only those samples for which sufficient numbers of known sex individuals were available (Irene Mound, Irene Mortuary, Santa Catalina de Guale, Santa Catalina de Guale de Amelia; none of the other precontact samples were large enough or complete enough for inclusion in these analyses). The results were not very informative, however, and the insignificant test statistics have been omitted for this rea-

Table 7.5. Guale maxillary univariate variance ratio confidence intervals

Measure	Comparison[a]	Ratio	num d.f.	den d.f.	CI 95%-
UI1BL	LPC vs SCDG	1.26	34	126	.77
	LPC vs AM	1.41	126	48	.85
	SCDG vs AM	1.77	34	48	.96
UCMD	LPC vs SCDG	1.22	115	97	.83
	LPC vs AM	1.06	115	43	.63
	SCDG vs AM	1.29	43	97	.79
UCBL	LPC vs SCDG	1.23	115	97	.82
	LPC vs AM	1.03	45	115	.63
	SCDG vs AM	1.09	45	97	.66
UP1MD	LPC vs SCDG	1.23	137	127	.87
	LPC vs AM	1.05	127	47	.64
	SCDG vs AM	1.29	137	47	.79
UP1BL	LPC vs SCDG	1.27	137	143	.98
	LPC vs AM	1.11	143	47	.68
	SCDG vs AM	1.41	137	47	.86
UM1MD	LPC vs AM	1.44	46	144	.92
UM1BL	LPC vs SCDG	1.16	143	188	.89
	LPC vs AM	1.03	188	50	.64
	SCDG vs AM	1.20	143	50	.74

Note: Ratio = variance ratio, num d.f. = numerator degrees of freedom, den d.f. = denominator degrees of freedom, CI 95%- = lower bound of 95% confidence interval.
[a] LPC = Aggregate Late Precontact, SCDG = Santa Catalina de Guale, AM = Santa Catalina de Guale de Amelia.

son (Table 7.7). For males, no significant tests resulted. Results for females were more informative with five significant Bartlett's tests and a single significant Levene's test. The results are consistent with those of the aggregate group analysis. For one measurement (UI1BL), Irene Mound exceeded Amelia Island in variability. Santa Catalina de Guale was more variable than Amelia Island for three measurements (LI2BL, LP1MD, LP1BL), more variable than Irene Mound for one measurement (LP1MD), and more variable than Irene Mortuary for two measurements (LP1MD, LP1BL). As with the combined-sex univariate results, phenotypic variability first increased in the mission period, followed by a subsequent decrease in the late mission period.

Table 7.6. Guale mandibular univariate variance ratio confidence intervals

Measure	Comparison[a]	Ratio	num d.f.	den d.f.	CI 95%-
LI2MD	LPC vs SCDG	1.89	80	87	1.23
LI2BL	LPC vs SCDG	1.16	61	123	.76
	LPC vs AM	1.45	123	50	.89
	SCDG vs AM	1.68	61	50	.98
LCMD	LPC vs SCDG	1.03	70	107	.68
	LPC vs AM	1.04	107	47	.63
	SCDG vs AM	1.08	70	47	.63
LCBL	LPC vs SCDG	1.12	137	61	.72
	LPC vs AM	1.02	137	54	.64
	SCDG vs AM	1.10	54	61	.65
LP1MD	LPC vs SCDG	2.14	136	130	1.52
	LPC vs AM	1.09	130	49	.67
	SCDG vs AM	2.32	136	49	1.41
LP1BL	LPC vs AM	1.42	145	51	.88
LM1MD	LPC vs AM	1.02	39	159	.65
LM1BL	LPC vs SCDG	1.00	186	141	.76

Note: Ratio = variance ratio, num d.f. = numerator degrees of freedom, den d.f. = denominator degrees of freedom, CI 95%- = lower bound of 95% confidence interval.

[a] LPC = Aggregate Late Precontact, SCDG = Santa Catalina de Guale, AM = Santa Catalina de Guale de Amelia.

As a final test of univariate variance differences, I repeated the sex-specific analyses using confidence intervals for the ratio of variances. The results for males are presented in Tables 7.8 and 7.9 for the maxilla and mandible, respectively. There were no significant results. The female analyses are presented in Tables 7.10 and 7.11 for the maxilla and mandible, respectively. The maxillary measurements produced no significant results. There were, however, four significant mandibular intervals, all involving an excess of variability for the Santa Catalina de Guale sample. Santa Catalina de Guale was more variable than Irene Mound, Irene Mortuary, and Amelia Island for LP1MD and was also more variable than Amelia Island for LI2BL. These data, therefore, support the Bartlett's and Levene's tests. In general, the univariate data indicate an initial increase in phenotypic variability in the early mission period followed by a

Table 7.7. Guale sex-specific univariate variance tests

Measure	Period[a]	n	sd	95%CI (sd)	Bartlett Stat (p-value)	Levene Stat (p-value)
Females						
UI1BL	IMD	21	.88	.63–1.42	15.62(.001)	1.16(.34)
	SCDG	6	.57	.32–1.88		
	AM	19	.38	.27–.62		
LI2BL	IMD	24	.44	.33–.70	18.83(.000)	.71(.55)
	IMT	10	.37	.22–.81		
	SCDG	8	1.00	.59–2.57		
	AM	25	.33	.24–.50		
LCBL	IMD	26	.35	.26–.54	18.17(.000)	4.84(.004)
	IMT	12	.99	.65–2.00		
	SCDG	9	.51	.31–1.21		
	AM	26	.58	.43–.88		
LP1MD	IMD	23	.49	.35–.76	70.73(.000)	1.34(.27)
	IMT	12	.51	.33–1.02		
	SCDG	17	2.39	1.65–4.14		
	AM	22	.50	.40–.88		
LP1BL	IMD	26	.49	.36–.75	23.55(.000)	2.08(.11)
	IMT	15	.37	.25–.67		
	SCDG	18	1.03	.72–1.75		
	AM	23	.46	.33–.71		

[a] IMD = Irene Mound, IMT = Irene Mortuary, SCDG = Santa Catalina de Guale, AM = Santa Catalina de Guale de Amelia.

subsequent decrease in the late mission period. This pattern was better defined for females.

Multivariate Analyses

ANOVA results for the combined-sex Van Valen's test statistic are presented in Table 7.12. The ANOVA p-value was not significant (p = .445), indicating equal variability across groups. Van Valen's test was repeated for the sex-specific samples using the same series of measurements. The result for males is presented in Table 7.13 and indicates no significant difference in between-sample variability. The ANOVA table for the female Van Valen's test is pre-

Table 7.8. Guale maxillary univariate variance ratio confidence intervals—males

Measure	Comparison[a]	Ratio	num d.f.	den d.f.	CI 95%-
UI1BL	IMD vs SCDG	2.53	2	17	.55
	IMD vs AM	1.18	17	21	.48
	SCDG vs AM	3.00	2	21	.68
UCMD	IMD vs SCDG	1.57	16	9	.42
	IMD vs AM	1.10	18	16	.40
	IMT vs SCDG	1.68	9	3	.77
	IMT vs AM	2.91	18	3	.21
	SCDG vs AM	1.73	18	9	.47
UCBL	IMD vs SCDG	1.26	20	9	.34
	IMD vs AM	1.03	20	20	.42
	IMT vs SCDG	1.29	5	9	.29
	IMT vs AM	1.02	20	5	.16
	SCDG vs AM	1.31	20	9	.36
UP1MD	IMD vs SCDG	1.00	22	8	.25
	IMD vs AM	1.46	22	22	.62
	SCDG vs AM	1.46	22	8	.37
UP1BL	IMD vs SCDG	1.71	23	9	.47
	IMD vs AM	1.14	21	23	.49
	IMT vs SCDG	1.27	3	9	.25
	IMT vs AM	1.14	21	3	.08
	SCDG vs AM	1.44	21	9	.39
UM1MD	IMD vs AM	2.64	13	15	.90
	IMT vs AM	2.98	12	5	.46
UM1BL	IMD vs SCDG	1.51	10	27	.58
	IMD vs AM	1.36	10	27	.53
	IMT vs SCDG	1.46	10	6	.27
	IMT vs AM	1.32	16	6	.25
	SCDG vs AM	1.11	16	10	.32

Note: Ratio = variance ratio, num d.f. = numerator degrees of freedom, den d.f. = denominator degrees of freedom, CI 95%- = lower bound of 95% confidence interval.

[a] IMD = Irene Mound, IMT = Irene Mortuary, SCDG = Santa Catalina de Guale, AM = Santa Catalina de Guale de Amelia.

Table 7.9. Guale mandibular univariate variance ratio confidence intervals—males

Measure	Comparison[a]	Ratio	num d.f.	den d.f.	CI 95%-
LI2MD	IMD vs SCDG	1.36	11	16	.45
	IMT vs SCDG	1.44	4	11	.34
LI2BL	IMD vs SCDG	2.20	25	9	.61
	IMD vs AM	1.61	25	9	.66
	IMT vs SCDG	1.50	9	7	.31
	IMT vs AM	2.05	19	7	.46
	SCDG vs AM	1.37	19	9	.37
LCMD	IMD vs SCDG	1.28	17	11	.39
	IMD vs AM	1.17	20	17	.45
	IMT vs SCDG	1.47	7	11	.39
	IMT vs AM	1.02	20	7	.23
	SCDG vs AM	1.50	20	11	.46
LCBL	IMD vs SCDG	1.10	12	24	.43
	IMD vs AM	1.06	24	25	.47
	IMT vs SCDG	1.95	12	18	.70
	IMT vs AM	1.66	25	18	.67
	SCDG vs AM	1.17	12	25	.47
LP1MD	IMD vs SCDG	1.45	15	26	.61
	IMD vs AM	1.08	26	23	.47
	IMT vs SCDG	1.03	15	10	.30
	IMT vs AM	1.51	10	23	.43
	SCDG vs AM	1.56	15	23	.63
LP1BL	IMD vs SCDG	1.08	14	29	.46
	IMD vs AM	1.19	29	24	.54
	IMT vs SCDG	1.02	14	11	.30
	IMT vs AM	1.26	12	24	.50
	SCDG vs AM	1.29	14	24	.52
LM1MD	IMD vs AM	1.73	8	24	.62
	IMT vs AM	1.27	8	9	.31
LM1BL	IMD vs AM	1.13	13	26	.46
	IMT vs AM	1.02	13	10	.28

Note: Ratio = variance ratio, num d.f. = numerator degrees of freedom, den d.f. = denominator degrees of freedom, CI 95%- = lower bound of 95% confidence interval.

[a] IMD = Irene Mound, IMT = Irene Mortuary, SCDG = Santa Catalina de Guale, AM = Santa Catalina de Guale de Amelia.

Table 7.10. Guale maxillary univariate variance ratio confidence intervals—females

Measure	Comparison[a]	Ratio	num d.f.	den d.f.	CI 95%-
UI1BL	IMD vs SCDG	1.54	20	5	.24
	IMD vs AM	2.32	20	18	.91
	SCDG vs AM	1.50	5	18	.44
UCMD	IMD vs SCDG	1.56	23	13	.54
	IMD vs AM	1.15	23	17	.45
	IMT vs SCDG	1.08	8	13	.32
	IMT vs AM	1.24	17	8	.31
	SCDG vs AM	1.35	17	13	.45
UCBL	IMD vs SCDG	1.09	26	16	.42
	IMD vs AM	1.02	26	17	.40
	IMT vs SCDG	1.12	16	11	.34
	IMT vs AM	1.20	17	11	.37
	SCDG vs AM	1.07	17	16	.39
UP1MD	IMD vs SCDG	2.07	17	22	.85
	IMD vs AM	1.56	20	22	.65
	SCDG vs AM	1.33	17	20	.53
UP1BL	IMD vs SCDG	1.15	15	26	.48
	IMD vs AM	1.31	26	21	.56
	IMT vs SCDG	2.73	15	16	.98
	IMT vs AM	1.82	21	16	.68
	SCDG vs AM	1.50	15	21	.59
UM1MD	IMD vs AM	1.30	18	22	.53
	IMT vs AM	1.40	18	11	.43
UM1BL	IMD vs SCDG	1.00	30	16	.39
	IMD vs AM	1.02	30	20	.17
	IMT vs SCDG	1.12	15	16	.40
	IMT vs AM	1.14	15	20	.44
	SCDG vs AM	1.02	16	20	.40

Note: Ratio = variance ratio, num d.f. = numerator degrees of freedom, den d.f. = denominator degrees of freedom, CI 95%- = lower bound of 95% confidence interval.

[a] IMD = Irene Mound, IMT = Irene Mortuary, SCDG = Santa Catalina de Guale, AM = Santa Catalina de Guale de Amelia.

Table 7.11. Guale mandibular univariate variance ratio confidence intervals—females

Measure	Comparison[a]	Ratio	num d.f.	den d.f.	CI 95%-
LI2MD	IMD vs SCDG	2.00	9	17	.67
LI2BL	IMD vs SCDG	2.27	7	23	.78
	IMD vs AM	1.33	23	24	.58
	IMT vs SCDG	2.70	7	9	.64
	IMT vs AM	1.12	9	24	.41
	SCDG vs AM	3.03	7	24	1.05
LCBL	IMD vs SCDG	1.46	8	25	.74
	IMD vs AM	1.66	25	25	.74
	IMT vs SCDG	1.94	11	8	.46
	IMT vs AM	1.71	11	25	.67
	SCDG vs AM	1.14	25	8	.29
LP1MD	IMD vs SCDG	4.87	16	22	1.97
	IMD vs AM	1.02	21	22	.43
	IMT vs SCDG	4.68	16	11	1.41
	IMT vs AM	1.02	11	21	.38
	SCDG vs AM	4.78	16	21	1.91
LP1BL	IMD vs SCDG	2.10	17	25	.89
	IMD vs AM	1.07	25	22	.46
	IMT vs SCDG	2.78	17	14	.96
	IMT vs AM	1.24	22	14	.44
	SCDG vs AM	2.24	17	22	.92

Note: Ratio = variance ratio, num d.f. = numerator degrees of freedom, den d.f. = denominator degrees of freedom, CI 95%- = lower bound of 95% confidence interval.
[a] IMD = Irene Mound, IMT = Irene Mortuary, SCDG = Santa Catalina de Guale, AM = Santa Catalina de Guale de Amelia.

Table 7.12. Van Valen's Test ANOVA for combined-sex Guale Province samples

n=49	Multiple R=.352		Squared Multiple R=.124		
Source	Sum of Squares	df	Mean Square	F	p-value
Sample	1.788	6	.298	.99	.445
Error	12.648	42	.301		

Table 7.13. Van Valen's Test ANOVA for Guale Province male samples

n=24	Multiple R=.544		Squared Multiple R=.296		
Source	Sum of Squares	df	Mean Square	F	p-value
Sample	.719	3	.240	2.80	.066
Error	1.712	20	.086		

Table 7.14. Van Valen's Test ANOVA for Guale Province female samples

n=24	Multiple R=.123		Squared Multiple R=.015		
Source	Sum of Squares	df	Mean Square	F	p-value
Sample	.075	3	.025	.10	.957
Error	4.885	20	.244		

sented in Table 7.14. As with the male results, the p-value for the test is insignificant at the 5% level (p = .957), suggesting no differences in variability between groups.

Results of the determinant ratio analysis are presented in Table 7.15. The boldface entries indicate the sample exhibiting the larger multivariate variability for each comparison. The pattern of variability differences is easy to interpret: when a mission period sample was compared to a precontact sample, the former was always found to be more variable. When the mission period samples were compared to each other, intrasite variability increased through time for males but decreased through time for females. It should be noted that the p-values for all male tests were not significant. For females, however, a single test was significant at the .05 level, and two additional tests were significant at the .10 level. Interpretation of the significant results is straightforward: both Santa Catalina de Guale and Santa Catalina de Guale de Amelia were more variable than the late precontact Irene Mound population. Comparison of contact period populations produced divergent results by sex. However, population genetic variability for females best demonstrates the general pattern of microevolutionary change: variability at first increases and then decreases during the later phases of the mission period, a result consistent with previous research (Griffin 1993; Griffin et al. 2001).

Matrix Decomposition Analysis

Because females demonstrate the microevolutionary trend most distinctly, the matrix decomposition analysis was performed for the females using the

Table 7.15. Determinant ratios for precontact to mission period variability comparisons

Group	ln(det2/det1)	p-value
Males		
Mound-**SCDG**	0.19	.441
Mortuary-**SCDG**	49.213	.148
Mound-**Amelia**	-0.16	.337
Mortuary-**Amelia**	-5.58	.341
SCDG-**Amelia**	-1.11	.176
Females		
Mound-**SCDG**	4.99	.093
Mortuary-**SCDG**	8.08	.079
Mound-**Amelia**	-0.53	.565
Mortuary-**Amelia**	-2.79	.040
SCDG-**Amelia**	4.13	.227

Note: Bold type signifies sample with higher variability.

Irene series and both mission period samples. Results for the tests of matrix correlations are presented for the raw and ranked distances in Table 7.16. The raw matrices do not transform the burial and genetic distances, whereas the ranked matrices convert the raw distances into ranked classes (see discussion in Stojanowski 2003). None of the matrix correlation p-values is significant, suggesting either a lack of kin structuring or hidden substructure in the cemeteries. Because it is reasonable to assume mission period cemeteries were kin-structured (see discussions in Jacobi 2000; McEwan 2001), application of the matrix decomposition model of Stojanowski (2003) seemed justified.

Sample sizes and decomposition proportions are presented in Table 7.17. As indicated, precontact samples have relatively large proportion values, while the mission period samples exhibit smaller values, suggesting that aggregation *increased* during the mission period, as expected. The confidence intervals for the pair-wise differences in proportions are presented in Table 7.18. Precontact samples do not differ significantly from each other; however, both the Santa Catalina de Guale and Amelia Island samples exhibit significantly smaller proportion values, indicating that the mission period samples are more subdivided than the precontact samples. In addition, Amelia Island is marginally more subdivided than the SCDG sample, suggesting continued aggregation of populations after the population nadir on the Georgia coast.

Table 7.16. Matrix correlation p-values for raw and ranked biological and burial distance matrices

Sample	Raw Data		Ranked data	
	r	p-value	r	p-value
Irene Mound	.02	.38	.06	.18
Irene Mortuary	–.07	.84	–.17	.48
Santa Catalina de Guale	.03	.48	.03	.59
Amelia Island	.11	.28	.10	.23

Table 7.17. Decomposition proportions for Guale females

Sample	N	NegRes	TotalRes	Proportion
Irene Mound	20	116	190	.61
Irene Mortuary	17	86	136	.63
SCDG	49	609	1176	.52
Amelia Island	19	76	171	.44

Note: N = sample size, NegRes= number of negative residuals, TotalRes = total number of residuals, Proportion = proportion of negative/total residuals.

Table 7.18. Confidence intervals for difference in proportions

	Irene Mound	Irene Mortuary	SCDG	Amelia
Irene Mound	0			
Irene Mortuary	(–.09, .13)	0		
SCDG	(.02, .16)[a]	(.01, .18)[a]	0	
Amelia	(.08, .40)[a]	(.08, .40)[a]	(.0003, .16)	0

[a] significant at p=.05 level

Demographic Collapse on the Georgia Coast

The results presented in this book are entirely consistent with those of Griffin (Griffin 1993; Griffin et al. 2001) and indicate that multiple forms of phenotypic data (cranial nonmetric, dental morphology, dental metric) provide similar interpretations, a fact that bolsters the explanatory arguments. My analysis differs from Griffin's in two ways. First, the segregation of sexes for some of the analyses allows greater resolution of historic period processes of accommodation and acculturation. Second, the implementation of the matrix decomposition model allows me to parcel the effects of admixture from population aggregation.

Aggregation and Demographic Collapse

The early mission period analyses (here more loosely defined, given the abandonment of Santa Catalina de Guale in the early 1680s; Larsen 1990) are consistent with the research model predicting that the 17th-century Guale were experiencing intensive population aggregation in the wake of increased morbidity, population stress, and demographic collapse, the last of which was counteracted by in-migration (partially voluntary) and population aggregation (of potential Spanish design) of disparate populations at the Santa Catalina mission. Phenotypic variability at Santa Catalina was greater than nine combined samples from the Georgia coast that date from A.D. 1200 to 1500. As with the Apalachee aggregate late precontact sample, the Guale LPC sample includes sites encompassing several hundred years (approximately 400) of burial activity along the Georgia coast, which is expected to maximize variability. Yet, Santa Catalina still demonstrates an increase in variability, despite its use for a more limited time period (approximately 74 years; Larsen, 1990). Therefore, documentation of increased variability is a robust result. Clearly populations using the Santa Catalina burial ground were of more diverse biological composition than those in earlier time periods. In other words, the burial catchment of Santa Catalina was much greater after contact.

There are many reasons to expect this pattern at Santa Catalina. Spanish presence is one possible vector of added variability since mestizaje seems very likely at this mission, despite scant historic evidence that this process actually occurred (Larsen 1990). It would be naive to assume that it did not occur, and Deagan does provide some circumstantial evidence for mestizaje. For example, in 1578 only six additional single women were requested of the Crown (Dunkle 1958:4), which Deagan interprets as evidence that soldiers were finding mates among the Guale (Deagan 1973:57). The effect of the presence of African slaves must also be considered, despite our inability to definitively place Africans in the historic record. Their presence is recorded, but their actions often are not.

In my opinion, however, the pattern more likely represents the effects of increasing contact and admixture between genetically distinct Native American populations (Worth 1995). This contact explains both the increase in variability and the decrease in the matrix decomposition proportion, suggesting that the number of distinct populations buried at Santa Catalina increased through time. I propose, therefore, the following cycle of events responsible for the accumulation of burials at Santa Catalina: (1) local populations were aggregated at the mission, a biological non-event because of the cultural, linguistic, and, to a lesser extent, political homogeneity recorded for the Guale subdistricts (Jones 1978); (2) epidemics and mission hardships lead to increased morbidity and mortality, out-migration further stressed these populations, and local population sizes declined rapidly by the terminal 16th century; (3) as population size declined, new groups were relocated to the mission via stage 1 or 2 congregación (see Worth 1995), other Christianized populations were locally aggregated with the Guale settlements (Type 1 reducción, e.g., the Mocama), or populations external to the mission system migrated to the region (Type 2 reducción, e.g., the Yamassee); (4) this cycle was repeated throughout the 17th century, resulting in an excess of genetic diversity documented here.

Evidence in support of this model is extensive and derives from various sources. Historical data suggest that pathogen transmission was a feature of early-16th-century Guale, beginning with slave raiding expeditions circa 1516 (Saunders 1992:140) and continuing throughout the period of Spanish exploration (Smith 1987). In fact, most of the epidemics documented in *La Florida*, which might be questionable for Apalachee (chapter 6), most certainly affected the coastal Guale and Timucua populations (Bushnell 1978a; Deagan 1973, 1990a, b; Hann 1986b, 1996; Milanich 1978, 1996, 1999; Sturtevant 1962; Thomas 1990b), owing in part to the extended period of European contact with this region (Brose 1984). Therefore, as opposed to the Apalachee, demographic collapse following disease epidemics seems a certainty. Historical sources do suggest that population sizes were declining, with consensus estimates of 6,000–8,000 in 1550, 1,200 in 1600, 670 in 1675, and 300 in 1702 (Boyd 1948; Milanich 1999; Worth 1995:13). By 1657, the population of Guale had been reduced to such a low number that the governor of Florida combined Guale and Timucua for administrative purposes (Worth 1995:13). I therefore expect the terminal population living at Santa Catalina to be much smaller and less diverse than the founder population.

In addition to population size decline, whether because of mortality, emigration, or both, a number of other processes more typical of the mission experience apply to the Guale. Repartimiento quotas were levied against the Guale, resulting in male transiency and fugitivism, which likely had identical effects on social organization as those proposed for the Apalachee. Guale females, who appear to have been more settled and less ephemeral in the Spanish system, were left without traditional mating partners, resulting not only in decreased birth rates but also in admixture with populations beyond normative prescriptions (Worth 1995:34). According to Worth, "Mission chiefs occasionally complained that most of the young males eligible for marriage lived or spent long periods of time in St. Augustine as wage laborers, leaving mission villages with a surfeit of unmarried young women" (Worth, 2001:18).

Changes in patterns of warfare and concomitant shifts in the political landscape also removed precontact barriers to travel and interaction. Most significant is the Spanish negotiated peace among warring tribes. As they did for the Apalachee, Yustaga, and various Muskogean populations north and west of Apalachee, the Spanish were successful negotiators of peace among Guale, Timucua, and Orista because it directly served their needs and fulfilled their charge to pacify and convert indigenous populations. Because it is known that Guale and Orista were at war at the time of contact (Lanning 1935:38) and that Guale and Mocama were at war (Worth 1995, 1998b), the resolution of these disputes led to more fluid population boundaries, resulting in an increase in biological interaction among these populations. That we see evidence for increased variability affirms positions that these groups were distinct before contact (chapter 3). Such accommodation was particularly acute after 1650, when Guale and Mocama become unified in the wake of demographic collapse (Worth 1995). Individual identities are maintained, but political divisions disappear and cultural homogenization ensues.

At the same time that Guale population sizes were declining and male transiency was increasing, numerous nonlocal groups appeared on the fringe of Guale territory. Chichimeco attacks began in 1661, the Yamassee appeared, disappeared, and reappeared beginning in the 1660s, and interaction with remnant eastern Timucua populations, in particular the Mocama, increased (Worth 1995), all providing opportunities for mate exchange previously unavailable. Although the severity of epidemics was maximized in Guale, the patterns of diversity at Santa Catalina de Guale, and the processes proposed to explain this diversity, are very similar to those operating during the tenure of San Luis's existence: contact, aggregation, and admixture. According to Worth,

> significant numbers of refugees from more distant regions ultimately settled within the mission system during the final decades of the 17th century (Worth, 1995, 1998b). Even though some of these immigrants remained only a short time within Spanish Florida, their presence within and adjacent to established mission communities undoubtedly resulted in at least some biological impact on local populations, particularly with regard to physical characteristics linked to ethnic origin. (Worth 2001:5)

The major difference between Santa Catalina and San Luis, however, is the bioarchaeological record of morbidity and stress in the former population. Although both cemeteries were extremely overcrowded with disturbed interments, skeletal indicators of poor health were less prevalent at San Luis. I propose that this lack reflects historical reality for the Apalachee, and I argue for a more subdued epidemic effect and a more integral out-migration effect in explaining temporal decreases in population size. However, there is little disparity between the historical, archaeological, bioarchaeological, and paleogenetic records for Santa Catalina. Historical claims of disease epidemics do manifest as increased morbidity and decreased health in the skeletal record.

Years of stable isotope investigation on the Georgia coast uniformly suggest an increase in the consumption of maize (less negative carbon signatures) and a decrease in the consumption of marine resources (less positive nitrogen signatures) after contact (Larsen et al. 2001; Schoeninger et al. 1990). There is little difference between early and late mission contexts; however, the transition to the mission period is marked. The effects of a maize-based diet on health can be dramatic owing to both amino acid deficiencies and the effects of reduced iron bioavailability, which leads to poor health (Larsen et al. 2001). Larsen concludes that "poor diets—and the associated poor nutrition—can exacerbate the effects of infection. Poor nutrition and infection have a synergistic relationship. . . . malnourished people are more susceptible to infection, and people with an infection have a worsened nutritional status" (Larsen et al. 2001:74–75). Such a position is well documented for the Guale.

Larsen and Harn (1994) examined the bone-specific prevalence of periostitis and found a significantly higher prevalence in the tibia and femur of contact period samples. They proposed that this resulted from population aggregation and the emergence of novel or more virulent forms of pathogens. Larsen and Sering (2000) and Schmidt (1993) investigated changes in the frequency of anemic indicators such as cribra orbitalia and porotic hyperostosis in the Guale populations. They reported significant increases in the prevalence of juvenile cribra orbitalia and porotic hyperostosis during the contact period, which may indicate the action of iron deficiency anemia or increased parasitic infection due to contamination of freshwater drinking sources (Schultz et al. 2001).

Skeletal indicators of morbidity were affirmed by dental pathology. Although Simpson (1990; Simpson et al. 2001) could not sample Santa Catalina for pathological striae of Retzius, hypoplastic defect frequencies were recorded for this mission (Larsen et al. 2001). Although Guale populations exhibited evidence for increased stress as measured by multiple parameters in comparison to interior Florida contemporaries, there was little temporal change in the number of affected individuals, the percentage of teeth affected, the mean hypoplasia frequency, or the mean hypoplasia width in comparing Santa Catalina to precontact local populations. Although there were some indications of increased metabolic disturbance in Santa Catalina, other variables were more ambiguous, suggesting a complex etiology to hypoplasia development.

Cross-sectional bone geometric properties suggest an increase in sedentism after contact, an increase in static workloads, perhaps reflecting an increase in body mass, and differential effects of long-range travel for some males involved in the repartimiento labor system (Larsen and Ruff 1994; Larsen et al. 1996; Ruff and Larsen 1990, 2001; Ruff et al. 1984). Pattern and prevalence of osteoarthritis indicated an increase in frequency of affected bones during the contact period. The most dramatic changes occurred in the vertebral joint surfaces with males exhibiting a higher frequency than females (Griffin and Larsen 1989; Larsen et al. 1996). This suggests an increase in the mechanical demands placed on the back, typical of lifting and hauling heavy objects as per repartimiento requirements.

Although bioarchaeological data generally support the position that morbidity was increasing through time, the effects of epidemics and emigration on population composition cannot be differentiated in my opinion. Guale defections were common and began well before the permanent establishment of missions in the region (Lanning 1935). Swanton (1922:91, 93) places emigrant Guale as far south as the Calusa chiefdom and as far north as South Carolina, and the problematic affiliation between Guale and later interior populations suggests a significant number of Guale were residing in central Georgia beyond the reach of the Spanish Crown.

If population sizes were declining, then we are once again met with the paradox of increased population genetic variability. Why don't we see a decline in variability for Santa Catalina? Accepting an estimate of 6,000–8,000 individuals for the precontact period (chapter 3), which declines to 1,200 individuals by the time the missionaries return to the coast after the 1597 rebellion (Lanning 1935:18), we expect to see evidence of this in terms of genetic variance measures. The answer lies in the resolution of the sampling design. We have no knowledge of the genetic variability of the founding population of Santa Catalina. We only know the aggregate level of variability of *all* populations that used Santa Catalina over the course of 74 years, and it was greater than in precontact times because more disparate populations, carrying more disparate alleles, are represented at the mission. We also know the level of variability in the population that abandoned Santa Catalina because these individuals are buried at the Amelia Island mission. Therefore, the effects of genetic drift are difficult to parcel with the archaeological resolution currently available. Aggregation and admixture effects take precedence as a result.

Prelude to Extinction

The contrast between the early and late mission period Guale samples well illustrates the sampling dilemma outlined for Apalachee in chapter 6. Because we know Santa Catalina and Amelia Island are the same church congregation represented by temporally discrete skeletal samples, the variability in the Amelia Island cemetery represents the terminal *living* population that relocated from Santa Catalina de Guale. In other words, the decrease in variability in the Amelia Island sample indicates that population size was decreasing throughout the time period that the Santa Catalina de Guale cemetery was in use, despite our inability to parcel these effects in the later burial population.

What I find more interesting is the documented increase in the number of resident biological lineages represented in the Amelia Island cemetery. Because this parameter (lineage number) increases diachronically, regardless of changes in total phenotypic variability, I can differentiate subpopulation and variance effects in a manner otherwise unavailable to previous researchers. For example, the transition from the precontact to early mission period is characterized by an increase in variability *and* in the number of lineages. Presumably, the approximately 74-year span of cemetery use at Santa Catalina indicates that those individuals buried just before abandonment were of a different generation than those buried in the early years of cemetery accumulation. It is probable that these individuals never met. That more lineages are represented at Santa Catalina than at Irene Mound suggests that periodic epidemics were affecting the population and that new groups were moving into the region in response. Although I cannot parcel the variance into an admixture and aggregation component, and likely some of the added variability is due to Spanish, African, or nonlocal Native American admixture, the suite of analyses offered in this chapter suggests that *some* of the variability does in fact result from historical processes of population turnover and replacement.

Considerable interpretive clarity is offered, however, for the Amelia Island population. Because variability decreased and the number of lineages continued to increase, the population history of the Guale can be characterized as the progressive aggregation of lineages that were relatively distinct but were decreasing rapidly in size and, as a result, in genetic variability. Such an interpretation is completely in line with ethnohistoric and archaeological models of the demographic transition for the coastal provinces (Lanning 1935; Larsen 1990; Worth 1995).

That these patterns were best demonstrated for the female subcomponent of the dataset further clarifies the changing role of the indigenous female throughout the contact period. And to clarify the above results, one needs only to substitute "matrilines" for "lineages" in the discussion. That is, the population history of the Guale can be characterized as the progressive aggregation of disparate and distinct matrilines (the core social unit) that were progressively decreasing in size. This is consistent with ethnohistoric data suggesting that the Guale were matrilineal, with documented increases in the number of female *cacicas* during the contact period, suggesting that females assumed more primary roles within their respective communities (Thomas 1990b), and with male transiency, which forced females to assume more prominent social roles in indigenous society.

The mission period Guale are represented by two temporally nonoverlapping samples known to be related to each other in an ancestor–descendant relationship. Santa Catalina de Guale represents the same political unit as the post-1686 Santa Catalina de Guale de Amelia Island sample. Late precontact samples are also diverse and include seven coastal Guale samples as well as two components from the Irene Mound site, which combined represent approximately 400 years of burial activity from a geographic area that is roughly coincident with the Guale polity.

Univariate and multivariate analyses present a complicated picture of changes in phenotypic variability during the historic period. The aggregate group tests indicated that variability increased through the early mission period and decreased in the late mission period. Sex-specific tests were also informative. For males, no significant differences in variability were evident. However, for females the univariate tests and the determinant ratio analysis suggest that variability increased in Santa Catalina de Guale in comparison to both the precontact and late mission period samples. Further refinement of this result through the matrix decomposition analysis indicates that a portion of the increased variability in Santa Catalina de Guale results from the consecutive aggregation of phenotypically distinct biological populations. The potential impact of Spanish or African admixture is also a consideration. The late-17th-century Guale experienced a significant population bottleneck, resulting in a loss of population genetic variability as represented at the Amelia Island mission. However, the matrix decomposition analysis also indicated that Amelia Island was more diverse than Santa Catalina in terms of the number of lineages represented. The population history of the Guale can be characterized as the progressive aggregation of distinct populations, each of which was individually decreasing in size owing to the effects of disease epidemics, repartimiento-related deaths, and selective out-migration throughout the 17th century. The remnant population that was finally removed to St. Augustine represented a small portion of the original inhabitants of the area. However, ethnic diversity likely remained high and increased in concert with population aggregation throughout the contact period.

8 Local and Global Histories

Culture Histories: A Synthesis

Temporal Comparisons: Early Mission Period

Consideration of the comparative experience of the Apalachee and Guale during the early phases of missionization produces discordant results. For the Apalachee, the transition from the precontact to mission period had fewer biological consequences. Overall variability in the population does not change significantly, suggesting neither a decline in population size nor increasing population aggregation or admixture. Stasis dominates. There is also little evidence for increased stress as pathological markers of metabolic disturbance actually decrease in frequency postcontact. Previously, researchers have interpreted this pattern as representing a shift in the nature of the stressors, from chronic to acute (Jones et al. 1991; Storey 1986). That is, people were dying before the diseases registered in the skeletal or dental tissues. However, this interpretation of the data, which is equally as valid (see Wood et al. 1992), implies a certain reticence, given the a priori assumption that Apalachee population size *must* have declined during the 16th and early 17th centuries. In my opinion the data may suggest otherwise. The genetic variance data do not indicate a decline in variability as expected under a drift model. Patale burial demography (Table 8.1) suggests a low burial rate of 4 individuals per year, and, assuming (yes, somewhat hypocritically) that a population of 500 individuals fell within Patale's burial catchment, results in a 0.8 percent burial rate per year. The lack of crowding and commingling at Patale is unique among known mission cemeteries (summary in McEwan 2001). In fact, if we compare burial disturbance data from Patale, Santa Catalina, and San Martín de Timucua (all three are roughly contemporaneous early mission period doctrinas), there is a perfect negative correlation between the percentage of undisturbed burials and the hy-

Table 8.1. Summary of mission burial demography

Mission	No.	% Dist.	No. Years	Burial Rate	Pop. Size	% Rate	Var. change
Patale	67	0	17	3.9	500	0.8	none
San Luis	900	50.4	48	18.7	1400	1.4	increase
SCDG	431	52.4	74	5.8	350	1.7	increase
Amelia	121	7.5	16	7.5	200	3.7	decrease

Notes: No. = sample size of burials, % Dist. = percentage of burials disturbed, No. Years = number of years mission cemetery was in use, Burial Rate = number of burials per year, Pop. Size = estimated living population size within the jurisdiction of each mission, % Rate = burial rate as a percentage of the estimated population size, Var. Change = change in phenotypic variability in comparison to antecedent population.

pothesized rate of population decline. That is, demographic collapse was earliest along the coast (Santa Catalina de Guale, 431 burials, 47.6 percent disturbed; Larsen 1990, 1993), later among the western Timucua (San Martín de Timucua, 23 excavated, 22 percent disturbed; Hoshower and Milanich 1993), and latest among the Apalachee (Patale, 67 individuals, 0 percent disturbed; Jones et al. 1991). Hann has noted a similar pattern:

> Thus, as one moved eastward along the mission trail from Apalachee's San Luis, the population declined steadily or one might say abruptly, once one crossed the Suwanne, reaching a low of 40 at Salamototo at the St. Johns River crossing. It is as if ripples from a pebble of contagion, dropped into the demographic puddle at St. Augustine, became increasingly attenuated in their impact as they moved out from the center. There is undoubtedly a temporal component in this as well, since regions with the smallest surviving populations had the longest sustained contact with the Spanish. (Hann 1990:10)

Parsimony suggests that disease, depopulation, and population aggregation were not a feature of the early contact period Apalachee experience.

Bioarchaeological data for Guale suggest that the mission experiences of the early contact period Apalachee and Guale were vastly different. First, comparison of relative average phenotypic variability[1] between precontact Guale and Apalachee indicates that the former (.58) was more variable than the latter (.52), which may reflect patterns of interaction, political organization, and warfare before contact. That is, the feared and respected Apalachee, living at higher population densities and isolated from neighboring groups by spatial buffer zones, may have had more limited biological interaction with regional populations. Variability at Santa Catalina de Guale (.71) increases the discrepancy in phenotypic variation and indicates an increase in variability during the early mission period. The interpretation offered for this result is consistent with historical constructions of population size and evidence for disease epidemics (Jones 1978; Lanning 1935; Worth 1995). The early contact period Guale were suffering more catastrophic losses due to disease and out-migration, and population aggregation began as a result. The suite of analyses performed was most consistent with the interpretation that multiple, discrete lineages were entering the burial record in succession. This is concordant with Spanish attempts to vivify their colony via Stage 1 and 2 congregación and then Type 1 and 2 reducción (Worth 1998a, b).

Additional bioarchaeological data suggest that disease was a significant component of population size decline in Guale. A majority of stress indicators increase in prevalence, which is consistent with dietary reconstructions and activity patterns suggesting a declining quality of life (Hutchinson and Larsen 2001; Larsen et al. 2001; Ruff and Larsen 2001). Once again, such patterns were not documented for the Apalachee living at Patale. Although extremely open to criticism, the burial structure at Santa Catalina de Guale (Table 8.1) similarly implicates disease as an avenue of demographic collapse. The 431 burials recovered from the cemetery represent 74 years of burial activity (Larsen 1993) or a rate of burial of 5.8 year. This is slightly higher than that observed for Patale. Given the degree of disturbance (not found at Patale) and the poor preservation of the Santa Catalina remains (similar to Patale), the recovery of 432 individuals likely underestimates the true number of burials at this mission, which is supported by the demographic analyses indicating infant under-enumeration (Russell et al. 1990). Therefore, the burial rate was likely higher than that reported. In addition, and equally as subject to the enumeration biases, the percentage of burials per total population of the mission was estimated as 1.7 percent per year based on an average population size of 350 individuals over three generations of burial accumulation. This rate of burial is double that calculated for Patale.

Temporal Comparisons: Late Mission Period

Comparison of data for the late mission period populations, here defined as those that postdate their respective antecedent populations, reveals continued disparity in the mission experience of Christian contemporaries. In Apalachee, mission San Luis mimics very closely the biological profile of Santa Catalina de Guale, which indicates a transition from the profile presented at Patale. Variability at San Luis is greater than that at Patale, which I interpret as resulting from population aggregation and genetic admixture with resident Spanish and perhaps nonlocal Native American populations. Comparison of the average variability of San Luis (.77) with Patale (.54) and Santa Catalina de Guale (.71) may suggest that admixture was more prevalent at San Luis than at Santa Catalina.

Diet and disease data for San Luis are limited. The single isotopic sample demonstrated an atypical carbon and nitrogen profile suggesting little maize and a high marine component in the diet (Hutchinson and Larsen 2001). Information on the prevalence of pathological striae of Retzius suggests limited childhood stress in this population (Simpson 2001). As the microscopy is based on very few samples (nine individuals), interpretation of the results should be done cautiously. Information on population size indicates that a decline was occurring, associated with historic evidence that epidemics did affect Apalachee after 1650 (Hann 1988) but had not reached the critical levels already achieved on the Georgia coast.

Burial demography at San Luis is also very similar to Santa Catalina de Guale (Table 8.1). San Luis contained an estimated 900 burials which accumulated over a period of 48 years, resulting in a burial rate of 19 burials per year. This is the highest burial rate of any mission, no doubt reflecting the large population the San Luis mission served. Assuming a population size of 1,400 individuals (McEwan 1991a), the burial rate as a percentage of the total population was 1.4 percent per year, which is comparable to the 1.7 percent documented for Santa Catalina and almost double that documented for Patale.

Populations living along the Georgia coast during this time period were experiencing catastrophic losses resulting from disease, slave raiding, and outmigration, and the remnant populations of the Guale chiefdoms congregated at the Amelia Island mission. The Amelia Island sample demonstrated a decrease in phenotypic variability in comparison to Santa Catalina de Guale (.53 and .71, respectively) and was much less variable than contemporaneous Apalachee populations. This must reflect the effects of genetic drift and indicates that the population that migrated from St. Catherine's to Amelia Island was considerably less diverse than the population that had founded the former mission. This conclusion is amply supported by historical documentation of population size that measured in the hundreds. Pathology and dietary analyses also indicate disparity in experience between Amelia Island and San Luis. The former demonstrated isotopic signatures typical of other historic contexts in which maize consumption was maximized and marine resource utilization was declining (Hutchinson and Larsen 2001). Pathological striae are ubiquitous, and measured morbidity is higher than at San Luis (Simpson 2001). Data for cribra orbitalia and porotic hyperostosis indicate an increasing prevalence of these conditions in comparison to Santa Catalina (Schultz et al. 2001).

Amelia Island had 121 burials that accumulated over a period of 16 years, resulting in a burial rate of 7.5 burials per year. This burial rate is higher than at Santa Catalina but lower than at San Luis, representing the disproportionate number of individuals in residence at each mission. Assuming a population size of 200 individuals on Amelia Island (Bushnell 1986; Saunders 1993), the burial rate as a percentage of the total population was 3.75 percent, nearly double that of San Luis and Santa Catalina. Clearly the rate of burial was accelerating among the remnant populations of the Georgia coast, a finding which is consistent with expectations based on bone and dental pathology.

A Synthetic Narrative

During the first decade of the 17th century, Christianity returned to the Guale Indians after nearly a decade of unrest in the wake of the 1597 uprising. Around the year 1607, the mission Santa Catalina de Guale was re-established on St. Catherine's Island. Although epidemics are recorded in coastal populations, and out-migration was significant after the 1597 rebellion, there is no evidence for a decline in population size in the paleogenetic data because the Santa Catalina cemetery does not represent the original founding population, but the founding population *and* others that congregated here subsequently. Although genetic drift is assumed to have been a significant factor that reduced genetic variance in concert with reductions in local population sizes, we currently lack the resolution to parcel such effects.

Life "under the bell" was clearly harsh for the Guale (see Larsen 2001). Population size declines caused stress on social institutions, which was compounded by labor quotas levied by the Spanish. Bioarchaeological data suggest that health declined throughout the period of Santa Catalina's use: dietary breadth decreased as maize came to dominate the diet. Most individuals became more sedentary while demands for physical labor increased. Decreased sanitation, increased infant diarrhea, and the effects of novel pathogens increased stress experience among the Guale.

By 1633, populations along the Georgia coast and in the St. Johns River valley were suffering catastrophic losses. In that year, Franciscan missionaries established the first of a dozen missions among the Apalachee after a 25-year period of intermittent contact. Although the reasons for missionary expansion are numerous, the demographic dilemma back east must have provided some inspiration for the friars. At about this time the San Pedro y San Pablo de Patale mission was established. As far as the data suggest, there was little change in Apalachee populations in the early contact period. Genetic variability was stable, which suggests limited loss of people between the precontact period and 1633. The rate of burial was low at Patale, health was apparently relatively unaffected by the missionaries' presence, and there was little evidence for population aggregation or population size declines by the time the mission was abandoned around 1650. The neat, orderly rows of interments speak to the lowest rate of burial among surveyed cemeteries. As the cemetery was unfilled, and cemetery overcrowding was apparently of little concern, the reasons for the relocation of the Patale congregation remain uncertain. At around the same time, one of the principal villages of the Apalachee also relocated. In 1656, San Luis de Talimali was re-established at the location at which it would remain throughout the mission period.

Back east, conditions deteriorated further as disease and depopulation continued to stress already weakened social institutions. The situation would worsen, however, after English establishment of key settlements farther north initiated shock waves of migrations of displaced peoples throughout the United States. Slave raiding attacks on the Guale began shortly after 1660 and would continue intermittently for the duration of the congregation's tenure on St. Catherine's Island (Worth 1995).

In general, the period between about 1650 and 1680 evoked a more homogenous response among Christian populations as issues of global significance took precedence. There is evidence for epidemics in both Guale and Apalachee, though I would argue of less severity in the latter. Displaced populations from inland and north appear at the fringe of Spanish territory. Demographic collapse was almost complete among eastern Timucua and Guale populations, and as a result labor demands increased for western Timucua and Apalachee at about this time. Christian Indians displaced after the mid-century interior uprisings became widely distributed throughout Spanish territory, while nonlocal, pagan populations appear within and among Apalachee settlements. Increasing numbers of shrinking populations continue to congregate on St. Catherine's Island as the rigors of mission life continued to exact a demographic toll.

In many ways, San Luis and Santa Catalina represent similar kinds of mission populations that were, in part, contemporaries. Although population density around San Luis was much higher than at Santa Catalina, the overall rate of burial disturbance was similar (~50 percent), the burial rate was similar (~1.5 percent), and both samples demonstrate an increase in population genetic variability in comparison to antecedent populations (Patale for San Luis and the precontact Guale for Santa Catalina). This mortuary profile reflects the processes of aggregation of nonlocal populations and subsequent admixture between local and immigrant groups. Different bioarchaeological disease profiles preface the composition of the populations that would eventually abandon these locations.

Sometime during the early 1680s, Santa Catalina was abandoned by a small but ethnically diverse group that represented a fraction of the precontact Guale genetically. This remnant, aggregate population resettled on Amelia Island. Although they were only in residence for 16 years, the rate of burial actually increased in comparison to Santa Catalina, 7.5 burials per year, which represented 3.7 percent of the living population. The rate of burial at Amelia Island was highest of all the missions, which corresponds with bioarchaeological data suggesting that health continued to decline in the coastal populations. One has the sense that a demographic Rubicon had been crossed in Guale, beyond which collapse accelerated because of the interaction between deleterious processes that fed off of one another. Population size, health, disease, work and activity levels, and indefensibility all interact in predictable and unfortunate ways.

That Apalachee was spared some of the catastrophic effects of colonization is attested to by final numbers. Approximately 6,000–8,000 Apalachee were living when the province was abandoned in the wake of English aggression, compared with the 200 remaining Guale. Although effects of disease and morbidity are uncertain, it seems likely that the expected repercussions of colonialism were accelerating at San Luis before it was abandoned. I highly suspect that genetic drift would have manifested in the San Luis population if a succeeding cemetery existed for analysis. The burgeoning conflict between England and Spain may have preempted future difficulties for the Apalachee.[2]

Broader Significance

In this book I have both questioned and affirmed received wisdom on the histories of Florida's indigenous populations. Specifically, I have linked regional variation in disease experience with observed differences in microevolutionary parameters for Apalachee and Guale. In terms of broader significance, these biological data speak on two related fronts to the fact that there was no singular response to contact.

First is the acrimonious debate about the effects of New World pandemics, the rate of demographic collapse, and the timing of demographic collapse. "High Counters"[3] promote the position that pandemics swept through New World indigenous communities, often far in advance of actual European contact (Dobyns 1966, 1983; Ramenofsky 1987; Smith 1987). Thus the earliest historical records are enumerating a population that has already suffered massive and catastrophic demographic losses. Individuals who prescribe to this position measure New World population size estimates in the millions, view pandemics as nearly universally affecting Native American communities, and propose rapid and catastrophic losses of 95 percent or greater. Others have debated the manner is which the population size estimates are achieved, citing questionable and abstract historical references to contagions and the circular manner in which epidemics are used to overinflate virgin soil population sizes (Baker and Kealhofer 1996; Hann 1988; Henige 1986, 1998; Milner 1980; Snow and Lanphear 1988, 1989).

A second, yet related, issue is the homogeneity of responses to colonialism. Mounting evidence from different areas suggests considerable regional diversity in the rate of population size decline and responses to such a process (Baker and Kealhofer 1996; Milner 1980, 1996; Ramenofsky et al. 2003; Stannard 1991; Walker 2001). Overextrapolation of models developed for one regional context to many different cultural groups inhabiting different environments and living under distinct social and political situations implies a monolithic view of the "colonist" and the "colonized." This perspective views Europeans as the power wielders and indigenous populations as the passive recipients of their future histories. In other words, the simplified version of European history—that is, Europeans arrive bringing disease, Native Americans lack immunity to the disease, massive epidemics decimate the native populations, and European domination of North America commences—assumes homogeneity in European New World policy, Native American response to this policy, and uniform underlying Native American social adaptations. There is little sense of agency in such a distillation.

The different biological profiles documented in this book are contrary to both the pandemicist and monolithicist positions. Health did not uniformly decline in *La Florida,* and epidemics did not uniformly ravage indigenous populations in the wake of initial contact. Responses to disease and colonialism were varied and localized. The documented pattern of phenotypic variance transformation was considerably different for Apalachee and Guale, suggesting that the two cannot be conflated into a pan–*La Florida* population for microevolutionary or bioarchaeological purposes. In other words, the distinct cultural and political histories of these populations did manifest in the population genetic signatures.

Despite my primary position that there *was* considerable regional variation in mission experience, I also believe that there was a prescribed pathway of change and that the differences observed for Guale and Apalachee are related more to the timing of events, most critically, the timing of population size decline. Consideration of the four missions included in this analysis reveals three distinct profiles (see Table 8.1). The first type of mission is represented by Patale with few burials, no burial disturbance (0 percent), a low rate of burial (3.9/year), a small percentage of burials as a percent of the living population (0.8 percent), a decrease in morbidity, and no change in phenotypic variability in comparison to antecedent populations. Patale represents a mission community *not* experiencing the typical contact period maladies. A Patale analogue does not exist for Guale either because we simply have not located such a mission or, more likely, because this stage of acculturation was brief or did not exist.

The second type of mission is represented by San Luis and Santa Catalina de Guale. Both missions had large numbers of burials, high rates of disturbance (~50 percent), higher rates of burial (18.75 and 7.5/year, respectively), and nearly equivalent burials rates as a percentage of the resident population (~1.5 percent). Both samples demonstrate an increase in phenotypic variation in comparison to antecedent populations. I propose that these cemeteries represent the period of active demographic transformation. Disease and out-migration were common, interaction between disparate cultural groups (both Native American and European) increased, congregation and aggregation of populations increased, and, as the social mechanisms preventing mate exchange broke down, admixture between diverse groups increased. That the pathological data for San Luis and Santa Catalina are not congruent implies some diversity in the mechanisms effecting these changes.

The third type of mission is represented by Amelia Island. This mission had an intermediate number of burials, an intermediate rate of disturbance (7.5 percent), a high rate of burial (7.5/year), the largest burial rate as a percentage of the living population (3.7 percent), increasing morbidity, and a decrease in phenotypic variability in comparison to the antecedent population. Such a profile implies acceleration of population size decline and represents the final stage in the process of demographic collapse.

Despite emphasis on the comparative disparity in biological signature profiles, the historical trajectories can be viewed as a continuum of reactions that differed in the timing of initial catalysts. The catalyst for these changes relates to the timing of the onset of population losses, whether owing to mortality or to out-migration. That is, the disparate histories of Apalachee and Guale may simply reflect the differential onset of the demographic transformation (Figure 8.1). Mission experience was maximally heterogenous before 1650, when Guale was in the midst of the demographic transition and Apalachee was largely unaffected. Experience homogenized after 1650 as represented by the similar bioarchaeological profiles of San Luis and Santa Catalina. Therefore, the localized nature of change can be seen as the end result of the differential experience of the postcontact but pre-mission (late 16th century) populations. The differential placement of the *x* in Figure 8.1 had ramifications for the Guale and Apalachee that were both immediate in effect and more time transgressive. The Guale are lost to history, whereas Apalachee are still a distinct identity today.

The Unanswered Question

Why was the early contact period experienced so differently by the Guale and Apalachee? I have presented the argument that the changes I have documented, as well as those documented by other bioarchaeologists (Larsen 2001), largely reflect differences in the timing of the demographic transition. However, there is no explanation for why the timing differed. Differential disease effects have been proposed, owing to the early contact along the coast as well as the nature

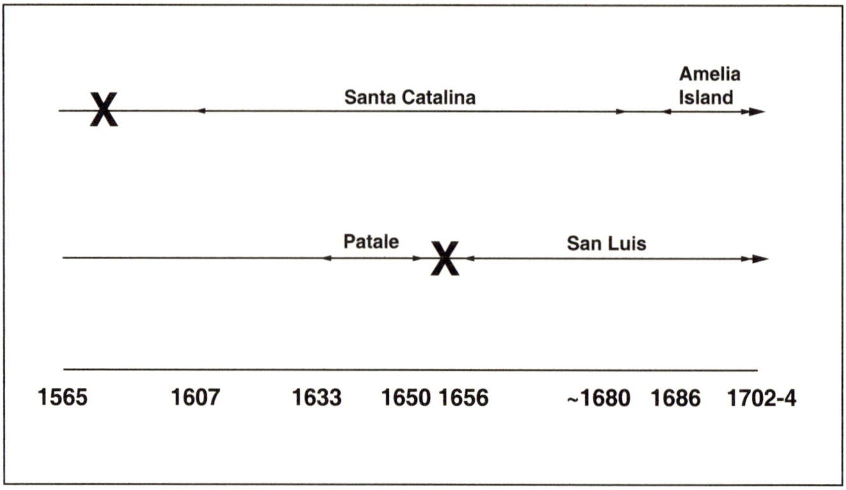

Figure 8.1. Graphic representation of the differential response proposed for Apalachee and Guale. Important dates are listed along the bottom with individual sample dates of use represented by the length of the arrow bars. The large 'x's' represent the approximate timing of the demographic transition within each province when population sizes began to decline significantly.

of the contact. Slave raiding expeditions along the Atlantic coast began early in the 1500s (Saunders 1992), and both French and Spanish efforts to claim this region resulted in short-lived 16th-century settlements ranging from Florida to South Carolina. As a point of contact for the Central American and Caribbean segments of the Spanish empire, the Atlantic coast may have been particularly prone to catastrophic disease incidents (Brose 1984; Deagan 1990b; Hann 1996; Saunders 1992). The demographic, historical, archaeological, and bioarchaeological data suggest that such was the case.

The isolated interior in which the Apalachee resided did not suffer these conditions until *after* the Spanish initiated peace between Apalachee and their neighbors (post-1650), at which time a period of lax sociopolitical circumscription (more migration) ensued. Although the 1528 Narváez expedition contacted the Apalachee and de Soto actually wintered there in the years 1539–40, the effects of introduced diseases were apparently minimal. Hann (1988) has proposed that the hostile nature of the interaction between European and indigenous communities best explains this anomaly. Indeed, the preponderance of data suggests that the Apalachee were given wide berth by their contemporaries. Linguistically distinct from neighboring groups and perennially at war with neighboring groups, the relatively populous province of Apalachee may have experienced a degree of isolation, represented in settlement distributions and accounts of geographic buffer zones physically isolating the polity. Ultimately, it may have been the success of their precontact sociopolitical strategy that saved Apalachee from the worst of the mission experience. Isolated and at war, things regionally may have seemed problematic. Internally, however, Apalachee was a large, vibrant, and populous nation and would remain comparatively so throughout much of the 17th century. Ranjel, a de Soto Chronicler, recorded:

> If their hands and noses were cut off they made no more account of it than if each of them had been a Mucius Scaevola of Rome. Not one of them, for fear of death, denied that he belonged to Apalache; and when they were taken and asked from whence they were they replied proudly: "From whence am I? I am an Indian of Apalache." And they gave one to understand that they would be insulted if they were thought to be of any other tribe than the Apalaches. (Ranjel in Bourne 1904:80)

This strong sense of identity may have served the Apalachee well throughout their history.

Conclusions

In this book I have attempted to tether multiple lines of evidence for the purpose of incorporating a paleogenetic perspective on the historical transformation of indigenous populations in 17th-century Spanish Florida. I approached contact period anthropology from four perspectives, each of which carries tenuous and untestable assumptions: (1) bioarchaeological indicators of health, metabolic stress, and morbidity; (2) basic burial demographic information specific to distinct cemeteries; (3) historical and ethnohistorical texts; and (4) paleogenetic analyses of population genetic variance transformations. Reasoned consideration of inherent weaknesses, and subsequent triangulation, of these data sources provides an opportunity to refine the regional and local histories of Florida's indigenous communities.

Evidence against the position that there were a uniform response and similar disease and morbidity experiences among Florida's converted is overwhelming. This is consistent with the work of others in this region (Larsen 2001) and the hemisphere in general (Larsen and Milner 1994). When synthesized, the preponderance of evidence suggests a regionally compartmentalized response to colonization rather than widespread and ubiquitous population size decline beginning immediately after, or even preceding, direct European contact. This book, therefore, becomes part of the growing literature in historical anthropology that dismisses colonialism as a monolithic institution (Baker and Kealhofer 1996; Henige 1986, 1998; Larsen and Milner 1994; Milner 1980, 1996). There is no biological response to colonialism in *La Florida* but rather biological *responses* to changing social conditions. Conflating Apalachee and Guale under the banner of *La Florida* ignores the distinct experiences and histories of these populations.

Adopting the data sources and perspectives of bioarchaeology has contributed to this regional scholarship in a manner unavailable to the historian and archaeologist. Bioarchaeology's broad consideration of both behavioral (activity patterns, disease, and diet) and evolutionary (gene flow and genetic drift) problems targets both biological and cultural aspects of the human condition. This places bioarchaeological approaches firmly within the center of a holistic anthropological research program. As a subdiscipline within biological anthropology, the dual emphasis on culture and evolution, and on history and process, well positions bioarchaeology to continue to make unique contributions to our understanding of the human species and its history. This book, I hope, has demonstrated this potential.

Notes

1. This is a simple measure based on the average of standard deviations for each sample, which corresponds nicely with univariate and multivariate variance analyses but is preferential because of its scalar quality.

2. This is not meant to imply that the killing and enslavement of some of the remaining Apalachee and their subsequent dispersal throughout the southeastern United States were not difficult.

3. After Henige (1998).

Appendix

The F test statistic is the ratio of variances for two samples. The test statistic is defined in Equation (2) as

$$s_x^2 / s_y^2$$

where s_x^2 is the larger variance and s_y^2 is the smaller variance. This statistic follows an F distribution with n_x-1 and n_y-1 degrees of freedom.

The multigroup equivalent of the F test is Bartlett's test, which is useful for variance comparisons among three or more groups. The test statistic is given as

$$\frac{(\sum v_i)\ln(\sum v_i S_i^2 / \sum v_1) - (\sum v_i \ln S_i^2)}{1 + \sum(1/v_i) - 1/\sum v_i / 3(k-1)}$$

where

$$S_i^2 = \sum_{j=1}^{n_i}(X_{ij} - \overline{X}_i)^2 / (n_i - 1)$$

and k = the number of samples, $v_i = n_i - 1$, X_{ij} = the *jth* observation for the *ith* sample and X^{bar} i = the sample mean. The test statistic is distributed as a χ^2 with $k-1$ degrees of freedom.

Levene's test is another multigroup variance comparison test. The test statistic is given as

$$\frac{(N-k)\sum n_i(V_i - V_{..})^2}{(k-1)\sum\sum(V_{ij} - V_{i.})^2}$$

where n_i = the individual sample sizes, N = the aggregate sample size, k = the number of samples, $V_{ij} = |X_{ij} - M_i|$ (where X_{ij} is the observation value for the *jth* individual *in sample i* and M_i = the sample median), V_i = individual observations, $V..$ = mean of V_{ij} for all ij, and $V_{i\cdot}$ = average V_{ij} value for the *ith* sample. Another way to envision Levene's test is as an ANOVA for k samples where the response variable equals the difference between measurement observations and the sample median.

Bonferroni confidence intervals are defined with the lower and upper bounds given as

$$[\sqrt{(n-1)/b} * s_i, \sqrt{(n-1)/a} * s_i]$$

where n is the sample size, s_i is the sample standard deviation, and a and b are chi-square values with $n-1$ degrees of freedom. The chi-squares for a and b are based on the individual error rate where a is half the individual error rate and b is 1-half the individual error rate. The individual error rate is determined by dividing the family-wise error rate by the number of samples for which confidence intervals are being estimated; the family-wise error rate is simply 1— the desired confidence level.

Endpoints for a confidence interval for the ratio of two variances is given as

$$[\frac{1}{F_{\alpha/2}(n-1,m-1)} * \frac{s_x^2}{s_y^2}, F_{\alpha/2}(m-1,n-1) * \frac{s_x^2}{s_y^2}$$

where $F_{\alpha/2}$ (n–1, m–1) is the F statistic for the desired confidence level with $n-1$ and $m-1$ degrees of freedom, s_x^2 is the sample variance for sample x and s_y^2 is the sample variance for sample y.

Van Valen's test is based on the distance of each observation from the within-group mean. The test statistic is given as

$$d_{ij} = \sqrt{\sum_{k=1}^{p} (X_{ijk} - \overline{X}_{jk})^2}$$

where X_{ijk} equals the value of X_k for the *ith* individual in sample j, and $X^{bar}{}_{jk}$ equals the mean of X_k for sample j. The resulting d_{ij} values can be compared between groups using a Student's t-test or ANOVA, depending on the number of groups being compared.

Bibliography

Albers, P. C. 1993. Symbiosis, merger, and war: Contrasting forms of intertribal relationship among historic plains Indians. *In* The political economy of North American Indians. J. H. Moore, ed. Pp. 94–134. Norman: University of Oklahoma Press.

Alvesalo, L. 1971. The influence of sex-chromosome genes on tooth size in man. Proceedings of the Finnish Dental Society 67:3–54.

Alvesalo, L., and M. Kari. 1977. Sizes of deciduous teeth in 47,XYY males. American Journal of Human Genetics 29:486–489.

Alvesalo, L., and P. Portin. 1980. 47,XYY males: Sex chromosomes and tooth size. American Journal of Human Genetics 32:955–959.

Alvesalo, L., and P. M. A. Tigerstedt. 1974. Heritabilities of human tooth dimensions. Hereditas 77:311–318.

Alvesalo, L., and J. Varrela. 1980. Permanent tooth sizes in 46,XY females. American Journal of Human Genetics 32:736–742.

Alvesalo, L., with R. H. Osborne and M. Kari. 1975. The 47,XYY male, Y chromosome, and tooth size. American Journal of Human Genetics 27:53–61.

Alvesalo, L., with E. Tammisalo and G. Townsend. 1991. Upper central incisor and canine tooth crown size in 47,XXY males. Journal of Dental Research 70:1057–1060.

Armelagos, G. J., and D. P. Van Gerven. 2003. A century of skeletal biology and paleopathology: Contrasts, contradictions, and conflicts. American Anthropologist 105:53–64.

Bachrach, F. H., and M. Young. 1927. A comparison of the degree of resemblance in dental characters shown in pairs of twins of identical and fraternal types. The British Dental Journal 48:1293–1304.

Bailit, H. L. 1975. Dental variation among populations. Dental Clinics of North America 19:125–139.

Baker, B. J., and L. Kealhofer, eds. 1996. Bioarchaeology of Native American adaptation in the Spanish borderlands. Gainesville: University Press of Florida.

Barcía, A. 1951. Chronological history of Florida. Gainesville: University of Florida Press.

Barrientos, B. 1567 [1965]. Pedro Menéndez de Avilés: Founder of Florida. Anthony Kerrigan, trans. Gainesville: University of Florida Press.

Beyer-Olsen, E. M. S., and V. Alexandersen. 1995. Sex assessment of Medieval Norwegian skeletons based on permanent tooth crown size. International Journal of Osteoarchaeology 5:274–281.

Bolton, H. E. 1917. The mission as a frontier institution in the Spanish-American colonies. American Historical Review 23:42–61.

Bolton, H. E. 1921. The Spanish borderlands: A chronicle of Old Florida and the Southwest. New Haven: Yale University Press.

Bourne, E. G. 1904. Narratives of the career of Hernando de Soto in the conquest of Florida. 2 vols. New York: A. S. Barnes.

Bowden, D. E. J., and D. H. Goose. 1969. Inheritance of tooth size in Liverpool families. Journal of Medical Genetics 6:55–58.

Boxer, C. F. 1975. Women in Iberian expansion overseas, 1415–1815. New York: Oxford University Press.

Boyd, M. F. 1948. Enumeration of Florida Spanish missions in 1675. Florida Historical Quarterly 27:181–188.

Boyd, M. F., with H. G. Smith and J. W. Griffin. 1951. Here they once stood: The tragic end of the Apalachee missions. Gainesville: University of Florida Press.

Boyd-Bowman, P. 1968. Regional origins of the Spanish colonists of America: 1540–1580. Buffalo Studies 1968:3–26.

Brace, C. L. 1991. What big teeth you had grandma! Human tooth size, past and present. *In* Advances in Dental Anthropology. M. A. Kelley and C. S. Larsen, eds. Pp. 33–58. New York: Wiley-Liss.

Braun, M. L. 1938. Likeness of dental arches of twins and of persons without blood relationship. Journal of the American Dental Association 25:1093–1098.

Brose, D. S. 1984. Mississippian period cultures in northwestern Florida. *In* Perspectives on Gulf Coast prehistory. D. D. Davis, ed. Pp. 165–197. Gainesville: University of Florida Press.

Buikstra, J. E., with S. R. Frankenberg and L. W. Konigsberg. 1990. Skeletal biological distance studies in American physical anthropology. American Journal of Physical Anthropology 82:1–7.

Buikstra, J. E., with L. W. Konigsberg and J. Bullington. 1986. Fertility and the development of agriculture in the prehistoric Midwest. American Antiquity 51:528–546.

Bushnell, A. T. 1978a. The Menéndez Marquez cattle barony at La Chua and the determinants of economic expansion in seventeenth-century Florida. Florida Historical Quarterly 56:407–431.

Bushnell, A. T. 1978b. "That demonic game": The campaign to stop Indian pelota playing in Spanish Florida, 1675–1684. The Americas 35:1–19.

Bushnell, A. T. 1979. Patricio de Hinachuba: Defender of the word of God, the Crown of the king, and the little children of Ivi-tachuco. American Indian Culture and Research Journal 3:1–21.

Bushnell, A. T. 1982. The king's coffer: Proprietors of the Spanish treasury, 1565–1702. Gainesville: University Presses of Florida.

Bushnell, A. T. 1986. Santa Maria in the written record. Miscellaneous Project Report Series 21. Florida State Museum, Department of Anthropology, Gainesville.

Bushnell, A. T. 1990. The sacramental imperative: Catholic ritual and Indian sedentism in the provinces of Florida. *In* Columbian consequences: 2. Archaeological and historical perspectives on the Spanish borderlands east. D. H. Thomas, ed. Pp. 475–490. Washington, DC: Smithsonian Institution Press.

Bushnell, A. T. 1994. Situado and sabana: Spain's support system for the presidio and mission provinces of Florida. Anthropological Papers of the American Museum of Natural History, No. 74. Athens: University of Georgia Press.

Caldwell, J., and C. McCann. 1941. Irene Mound site, Chatham County, Georgia. Athens: University of Georgia Press.

Cheverud J. M., and J. E. Buikstra. 1981a. Quantitative genetics of skeletal nonmetric traits in the rhesus macaques of Cayo Santiago. I. Single trait heritabilities. American Journal of Physical Anthropology 54:43–49.

Cheverud, J. M., and J. E. Buikstra. 1981b. Quantitative genetics of skeletal nonmetric traits in the rhesus macaques on Cayo Santiago. II. Phenotypic, genetic, and environmental correlations between traits. American Journal of Physical Anthropology 54:51–58.

Cheverud, J. M., and J. E. Buikstra. 1982. Quantitative genetics of skeletal nonmetric traits in the rhesus macaques of Cayo Santiago. III. Relative heritability of skeletal nonmetric and metric traits. American Journal of Physical Anthropology 59:151–155.

Christian, J. C. 1979. Testing twin means and estimating genetic variance: Basic methodology for the analysis of quantitative twin data. Acta Geneticae Medicae et Gemellologiae 28:35–40.

Christian, J. C., and J. A. Norton. 1977. A proposed test of the differences between the means of monozygotic and dizygotic twins. Acta Geneticae Medicae et Gemellologiae 26:49–53.

Christian, J. C., with K. W. Kang and J. A. Norton. 1974. Choice of an estimate of genetic variance from twin data. American Journal of Human Genetics 26:154–161.

Cohen, J. T., with C. P. Oliver and S. Bernick. 1942. Dental studies of triplets. Journal of Dental Research 21:233–239.

Cook, F. 1980. Aboriginal mortality on the Georgia coast during the early historic period. South Carolina Antiquities 12:36–42.

Cordell, A. S. 2002. Continuity and change in Apalachee pottery manufacture. Historical Archaeology 36:36–54.

Corruccini, R. S., and R. H. Y. Potter. 1980. Genetic analysis of occlusal variation in twins. American Journal of Orthodontics 78:140–154.

Covington, J. W. 1964. The Apalachee Indians move west. The Florida Anthropologist 17: 221–225.

Covington, J. W. 1968. Stuart's Town, the Yamassee Indians and Spanish Florida. The Florida Anthropologist 21:8–13.

Dahlberg, A. A. 1951. The dentition of the American Indian. *In* Papers on the physical anthropology of the American Indian. W. S. Laughlin, ed. Pp. 138–176. New York: Viking Fund.

Deagan, K. A. 1973. Mestizaje in colonial St. Augustine. Ethnohistory 20:53–65.

Deagan, K. A. 1978. Cultures in transition: Fusion and assimilation among the eastern Timucua. *In* Tacachale: Essays on the Indians of Florida and southeastern Georgia during the historic period. J. Milanich and S. Proctor, eds. Pp. 89–119. Gainesville: University Presses of Florida.

Deagan, K. 1983. Spanish St. Augustine: The archaeology of a colonial Creole community. New York: Academic Press.

Deagan, K. A. 1985. Spanish-Indian interaction in sixteenth century Florida and Hispaniola. *In* Cultures in contact. W. W. Fitzhugh, ed. Pp. 281–318. Washington, DC: Smithsonian Institution Press.

Deagan, K. A. 1990a. Accommodation and resistance: The process and impact of Spanish colonization in the Southeast. *In* Columbian consequences: 2. Archaeological and historical perspectives on the Spanish borderlands east. D. H. Thomas, ed. Pp. 297–314. Washington, DC: Smithsonian Institution Press.

Deagan, K. A. 1990b. Sixteenth-century Spanish-American colonization in the southeastern United States and the Caribbean. *In* Columbian consequences: 2. Archaeological and historical perspectives on the Spanish borderlands east. D. H. Thomas, ed. Pp. 225–250. Washington, DC: Smithsonian Institution Press.

Dempsey, P. J., and G. C. Townsend. 2001. Genetic and environmental contributions to variation in human tooth size. Heredity 86:685–693.

Dempsey, P. J., with G. C. Townsend, N. G. Martin, and M. C. Neale. 1995. Genetic covariance structure of incisor crown size in twins. Journal of Dental Research 74:1389–1398.

DePratter, C. B. 1991. Late prehistoric and early historic chiefdoms in the southeastern United States. New York: Garland Publishing, Inc.

Devor, E. J., with M. McGue, M. H. Crawford, and P. M. Lin. 1985. Transmissible and nontransmissible components of anthropometric variation in the Alexanderwohl Mennonites. II. Resolution by path analysis. American Journal of Physical Anthropology 69: 83–92.

Dittmar, M., with V. Galera, M. Garralda, and M. Smets. 1998. Permanent dentition in Santa Maria de Hito populations (Spain): Tooth crown diameters and microevolution in the Iberian peninsula. Homo 49:55–73.

Dobyns, H. F. 1966. Estimating aboriginal American population 1: An appraisal of techniques with a new hemispheric estimate. Current Anthropology 7:395–449.

Dobyns, H. F. 1983. Their number become thinned: Native American population dynamics in eastern North America. Knoxville: University of Tennessee Press.

Dobyns, H. F. 1991. The invasion of Florida: Disease and the Indians of Florida. *In* Spanish pathways in Florida: 1492–1992. A. L. Henderson and G. R. Mormino, eds. Pp. 58–77. Sarasota: Pineapple Press, Inc.

Dobyns, H., and W. Swagerty. 1980. Timucuan population in the 1560's. Paper presented at the Southeastern Archaeological Conference, New Orleans.

Dockrell, R. B. 1956. Tooth size in Irish (Aran Island) families. Journal of the European Orthodontic Society 32:200–216.

Doran, G. A., and L. Freedman. 1974. Metrical features of the dentition and arches of populations from Goroka and Lufa, Papua, New Guinea. Human Biology 46:583–594.

Dunkle, J. R. 1958. Population change as an element in the historical geography of St. Augustine. Florida Historical Quarterly 37:3–32.

Ebeling, C. F., with B. Ingervall, B. Hedegärd, and T. Lewin. 1973. Secular changes in tooth size in Swedish men. Acta Odontologica Scandinavica 31:140–147.

Edgington, E. S. 1987. Randomization tests. New York: Marcel Drekker.

Ehrmann, W. W. 1940. The Timucua Indians of sixteenth century Florida. Florida Historical Quarterly 18:168–191.

El Nofely, A., and W. A. Tawfik. 1995. On inheritance of permanent teeth crown size in a Middle-Eastern population. Homo 46:51–62.

Elvas, Gentleman of. 1907. The narrative of the expedition of Hernando De Soto, by the gentleman of Elvas. New York: Charles Scribner's Sons.

Ezzo, J. A., with C. S. Larsen and J. H. Burton. 1995. Elemental signatures of human diets from the Georgia Bight. American Journal of Physical Anthropology 98:471–481.

Falconer, D. S. 1981. Introduction to quantitative genetics. London: Longman.

Falconer, D. S., and T. F. C. Mackay. 1996. Introduction to quantitative genetics. 4th edition. New York: Prentice-Hall.

Falk, D., and R. Corruccini. 1982. Efficacy of cranial versus dental measurements for separating human populations. American Journal of Physical Anthropology 57:123–127.

Feldman, M. W., and R. C. Lewontin. 1975. The heritability hang-up. Science 190:1163–1168.

Filipsson, R., and L. Goldson. 1963. Correlation between tooth width, width of the head, length of the head, and stature. Acta Odontologica Scandinavica 21:359–365.

Filipsson, R., with J. Lindsten and S. Almquist. 1965. Time of eruption of the permanent teeth, cephalometric and tooth measurement and sulphation factor activity in 45 patients with Turner's syndrome with different types of X-chromosome aberration. Acta Endocrinology 48:91–113.

Fresia, A. E., and C. B. Ruff. 1987. Temporal decline in bilateral asymmetry of the upper limb on the Georgia coast. American Journal of Physical Anthropology 72:199.

Fresia, A. E., with C. B. Ruff and C. S. Larsen. 1990. Temporal decline in bilateral asymmetry of the upper limb on the Georgia coast. *In* The archaeology of mission Santa Catalina de Guale: 2. Biocultural interpretations of a population in transition. C. S. Larsen, ed. Pp. 121–132. New York: Anthropological Papers of the American Museum of Natural History no. 68.

Gannon, M. V. 1983. The cross in the sand: The early Catholic church in Florida, 1513–1870. Gainesville: University Presses of Florida.

Gardner, W. M. 1966. The Waddells Mill Pond site. The Florida Anthropologist 19:43–60.

Garn, S. M., with A. B. Lewis and R. S. Kerewsky. 1965a. Genetic, nutritional, and maturational correlates of dental development. Journal of Dental Research 1 (supplement): 228–243.

Garn, S. M., with A. B. Lewis and R. S. Kerewsky. 1965b. Size interrelationships of the mesial and distal teeth. Journal of Dental Research 44:350–354.

Garn, S. M., with A. B. Lewis and R. S. Kerewsky. 1965c. Relationship between buccolingual and mesiodistal tooth diameters. Journal of Dental Research 44:495.

Garn, S. M., with A. B. Lewis and R. S. Kerewsky. 1967a. Buccolingual size asymmetry and its developmental meaning. The Angle Orthodontist 37:186–193.

Garn, S. M., with A. B. Lewis and R. S. Kerewsky. 1967b. Sex difference in tooth shape. Journal of Dental Research 46:1470.

Garn, S. M., with A. B. Lewis and R. S. Kerewsky. 1968. Relationship between buccolingual and mesiodistal tooth diameters. Journal of Dental Research 47:495.

Garn, S. M., with A. B. Lewis and A. J. Wallenga. 1968. Two-generation confirmation of crown-size profile pattern. Journal of Dental Research 47:1190.

Garn, S. M., with A. B. Lewis, D. R. Swindler, and R. S. Kerewsky. 1967c. Genetic control of sexual dimorphism in tooth size. Journal of Dental Research 5 (supplement):963–972.

Geiger, M. 1937. The Franciscan conquest of Florida (1573–1618). Ph.D. dissertation, Catholic University of America.

Gillett, H. W. 1927. Contacts between archaeological and dental research. American Anthropologist 29:291–295.

Goggin, J. M., and W. C. Sturtevant. 1964. The Calusa: A stratified, nonagricultural society. *In* Explorations in cultural anthropology: Essays in honor of George Peter Murdock. W. H. Goodenough, ed. Pp. 179–219. New York: McGraw Hill.

Goose, D. H. 1963. Dental measurement: An assessment of its value in anthropological studies. *In* Dental anthropology. D. R. Brothwell, ed. Pp. 125–148. New York: Pergamon Press.

Goose, D. H. 1971. The inheritance of tooth size in British families. *In* Dental morphology and evolution. A. A. Dahlberg, ed. Pp. 144–149. Chicago: University of Chicago Press.

Granberry, J. 1993. A grammar and dictionary of the Timucua language. Tuscaloosa: University of Alabama Press.

Grant, D. A., with I. B. Stern and F. G. Everett. 1979. Periodontics. St. Louis: Mosby.

Gravlee, C. C., with H. R. Bernard and W. R. Leonard. 2003. Heredity, environment, and cranial form: A re-analysis of Boas's immigrant data. American Anthropologist 105:125–138.

Green, P. E. 1976. Mathematical tools for applied multivariate analysis. New York: Academic Press.

Griffin, J. W. 1990. Changing perspectives on the Spanish missions of La Florida. *In* Columbian consequences: 2. Archaeological and historical perspectives on the Spanish borderlands east. D. H. Thomas, ed. Pp. 399–408. Washington, DC: Smithsonian Institution Press.

Griffin, M. C. 1989. Dental variation of native populations from northern Spanish Florida. M.A. thesis, Department of Anthropology, Northern Illinois University.

Griffin, M. C. 1993. Morphological variation of the late precontact and contact period Guale. Ph.D. dissertation, Department of Anthropology, Purdue University.

Griffin, M. C., and C. S. Larsen. 1989. Patterns in osteoarthritis: A case study from the prehistoric and historic southeastern U.S. Atlantic coast. American Journal of Physical Anthropology 78:232.

Griffin, M. C., and A. M. Nelson. 1996. Adult and deciduous dental morphology in four U.S. Amerindian population samples: A comparative study. American Journal of Physical Anthropology Supplement 22:175.

Griffin, M. C., with P. M. Lambert and E. Monahan-Driscoll. 2001. Biological relationships and population history of native peoples in Spanish Florida and the American southeast. *In* Bioarchaeology of Spanish Florida: The impact of colonialism. C. S. Larsen, ed. Pp. 226–273. Gainesville: University Press of Florida.

Grön, M., and L. Alvesalo. 1995. Dental occlusion and arch size and shape in 46,XY females. *In* Proceedings of the 10th international symposium on dental morphology. R. J. Radlanski and H. Renz, eds. P. 123. Berlin.

Haas, M. R. 1971. Southeastern Indian linguistics. *In* Red, white, and black: Symposium on Indians in the Old South. C. M. Hudson, ed. Pp. 44–54. Athens: University of Georgia Press.

Hakluyt, R. 1903 [1600]. The principal navigations, voyages, traffiques and discoveries of the English nation, vol. 9. Reprinted by James MacLehose, Glasgow.

Hanke, L. 1964. The first social experiments in America. Gloucester, MA: Peter Smith.

Hann, J. H. 1986a. Spanish translations. Florida Archaeology no. 2. Tallahassee: Florida Bureau of Archaeological Research.

Hann, J. H. 1986b. Demographic patterns and changes in mid-seventeenth century Timucua and Apalachee. Florida Historical Quarterly 64:371–392.

Hann, J. H. 1987. Twilight of the Mocamo and Guale aborigines as portrayed in the 1695 Spanish visitation. Florida Historical Quarterly 66:1–24.

Hann, J. H. 1988. Apalachee: The land between the rivers. Gainesville: University Presses of Florida.

Hann, J. H. 1990. De Soto, Dobyns, and demography in Western Timucua. The Florida Anthropologist 43:3–12.
Hann, J. H. 1991. Father Juan de Paiva: Spanish friar of colonial Florida. *In* Spanish pathways in Florida: 1492–1992. A. L. Henderson and G. R. Mormino, eds. Pp. 140–146. Sarasota: Pineapple Press, Inc.
Hann, J. H. 1993. Visitations and revolts in Florida, 1656–1695. Florida Archaeology no. 7. Tallahassee: Florida Bureau of Archaeological Research.
Hann, J. H. 1996. A history of the Timucua Indians and missions. Gainesville: University Press of Florida.
Hann, J. H., and B. G. McEwan. 1998. The Apalachee Indians and mission San Luis. Gainesville: University Press of Florida.
Hardin, K. W. 1986. The Santa Maria mission project. The Florida Anthropologist 39:75–83.
Harris, E. F., and R. J. Smith. 1980. A study of occlusion and widths in families. American Journal of Orthodontics 78:155–163.
Harris, E. F., and R. J. Smith. 1982. Occlusion and arch size in families: A principal components analysis. The Angle Orthodontist 52:135–143.
Hartl, D. L., and A. G. Clark. 1987. Principles of population genetics. Sunderland, MA: Sinauer Associates, Inc.
Harzer, W. 1995. Size relationships and interactions between crown diameters in twins and family members. *In* Proceedings of the 10th international symposium on dental morphology. R. J. Radlanski and H. Renz, eds. Pp. 124–135. Berlin.
Hauser, G., and G. F. De Stefano. 1989. Epigenetic variants of the human skull. Stuttgart: Schweizerbart.
Henderson, A. M., and R. S. Corruccini. 1976. Relationship between tooth size and body size in American Blacks. Journal of Dental Research 55:94–96.
Henige, D. P. 1986. Primary source by primary source? On the role of epidemics in New World depopulation. Ethnohistory 33:293–312.
Henige, D. P. 1998. Numbers from nowhere. Norman: University of Oklahoma Press.
Hillson, S. 1996. Dental anthropology. Cambridge: Cambridge University Press.
Hoshower, L. M., and J. T. Milanich. 1993. Excavations in the Fig Springs mission burial area. *In* The Spanish missions of *La Florida*. B. G McEwan, ed. Pp. 217–243. Gainesville: University Press of Florida.
Houghton, P. 1996. People of the great ocean. Cambridge: Cambridge University Press.
Hu, J. R., with A. Nakasima and Y. Takahama. 1991. Heritability of dental arch dimensions in humans. Journal of Craniofacial Genetics and Developmental Biology 11:165–169.
Hudson, C. M. 1976. Southeastern Indians. Knoxville: University of Tennessee Press.
Hulse, F. S. 1941. The people who lived at Irene. *In* Irene Mound site, Chatham County, Georgia. J. Caldwell and C. McCann, eds. Pp. 57–68. Athens: University of Georgia Press.
Hussey, R. D. 1932. Text of the Laws of Burgos (1512–1513) concerning the treatment of the Indians. Hispanic American Historical Review 12:301–326.
Hutchinson, D. L. 1986. Enamel hypoplasia: Biological stress and lifeway on St. Catherines Island, Georgia. M.A. thesis, Department of Anthropology, Northern Illinois University.
Hutchinson, D. L. 1990. Postcontact biocultural change: Mortuary site evidence. *In* Columbian consequences: 2. Archaeological and historical perspectives on the Spanish borderlands east. D. H. Thomas ed. Pp. 61–70. Washington, DC: Smithsonian Institution Press.

Hutchinson, D. L., and C. S. Larsen. 1988. Determination of stress duration from linear enamel hypoplasias: A case study from St. Catherines Island, Georgia. Human Biology 60:93–110.

Hutchinson D. L., and C. S. Larsen. 1990. Stress and lifeway change: The evidence from enamel hypoplasias. *In* The archaeology of mission Santa Catalina de Guale: 2. Biocultural interpretations of a population in transition. C. S. Larsen, ed. Pp. 50–65. New York: Anthropological Papers of the American Museum of Natural History no. 68.

Hutchinson, D. L., and C. S. Larsen. 2001. Enamel hypoplasia and stress in La Florida. *In* Bioarchaeology of Spanish Florida: The impact of colonialism. C. S. Larsen, ed. Pp. 181–206. Gainesville: University Press of Florida.

Hutchinson, D. L., with C. S. Larsen, L. Norr, and M. J. Schoeninger. 2000. Agricultural melodies and alternative harmonies in Florida and Georgia. *In* Bioarchaeological studies of life in the age of agriculture: A view from the Southeast. P. M. Lambert, ed. Pp. 96–115. Tuscaloosa: University of Alabama Press.

Hutchinson, D. L., with C. S. Larsen, M. J. Schoeninger, and L. Norr. 1998. Regional variation in the pattern of maize adoption and use in Florida and Georgia. American Antiquity 63:397–416.

Jacobi, K. P. 2000. Last rites for the Tipu Maya: Genetic structuring in a colonial cemetery. Tuscaloosa: University of Alabama Press.

Johnson, K. W. 1991. The Utina and the Potano peoples of northern Florida: Changing settlement systems in the Spanish colonial period. Ph.D. dissertation, Department of Anthropology, University of Florida.

Jones, B.C. 1982. Southern cult manifestations at the Lake Jackson site, Leon County, Florida: Salvage excavation of mound 3. Midcontinental Journal of Archaeology 7:3–44.

Jones, B.C., and G. N. Shapiro. 1990. Nine mission sites in Apalachee. *In* Columbian consequences: 2. Archaeological and historical perspectives on the Spanish borderlands east. D. H. Thomas, ed. Pp. 491–510. Washington, DC: Smithsonian Institution Press.

Jones, B.C., with R. Storey and R. Widmer R. 1991. The Patale cemetery: Evidence concerning the Apalachee mission mortuary complex. *In* San Pedro y San Pablo de Patale: A seventeenth-century Spanish mission in Leon County, Florida. Florida Archaeology no. 5. B. S. Jones, J. Hann, and J. F. Scarry, eds. Pp. 109–125. Tallahassee: Florida Bureau of Archaeological Research.

Jones, G. D. 1978. The ethnohistory of the Guale coast through 1684. *In* The anthropology of St. Catherines Island. 1. Natural and cultural history. Anthropological Papers 55, pt. 2. D. H. Thomas, G. D. Jones, R. S. Durham, and C. S. Larsen, eds. Pp. 178–210. New York: American Museum of Natural History.

Jones, G. D. 1980. Guale Indians of the southeastern United States coast. *In* Excursions in southeastern geology: The archaeology-geology of the Georgia coast. J. D. Howard, C. B. DePratter, and R. W. Frey, eds. Pp. 215–224. Atlanta: Geological Society of America.

Jorde, L. B. 1980. The genetic structure of subdivided human populations: A review. *In* Current developments in anthropological genetics, vol. 1: Theory. J. H. Mielke and M. H. Crawford, eds. Pp. 133–206. New York: Plenum Press.

Kari, M., with L. Alvesalo and K. Manninen. 1980. Sizes of deciduous teeth in 45,X females. Journal of Dental Research 59:1382–1385.

Kieser, J. A. 1985. An odontometric analysis of the early Griqua dentition. Anthropologischer Anzeiger 20:103–107.

Kieser, J. A. 1990. Human adult odontometrics: The study of variation in adult tooth size. New York: Cambridge University Press.

Kieser, J. A., with H. T. Groeneveld and C. B. Preston. 1985a. Age related tooth size variation in the Lengua Indians of Paraguay. Zeitshcrift für Morphologie und Anthropologie 75:341–348.

Kieser, J. A., with H. T. Groeneveld and C. B. Preston. 1985b. Craniofacial correlates of attrition. The Angle Orthodontist 55:329–335.

Kieser, J. A., with H. T. Groeneveld and C. B. Preston. 1985c. Patterns of dental wear in the Lengua Indians of Paraguay. American Journal of Physical Anthropology 66:21–29.

King, J. 1984. Ceramic variability in 17th century St. Augustine, Florida. Historical Archaeology 18:76–82.

King, L., with E. F. Harris and E. A. Tolley. 1993. Heritability of cephalometric and occlusal variables as assessed from siblings with overt malocclusions. American Journal of Orthodontics and Dentofacial Orthopedics 104:121–131.

Kolakowski, D., and H. L. Bailit. 1981. A differential environmental effect on human anterior tooth size. American Journal of Physical Anthropology 54:377–381.

Konigsberg, L. W. 1988. Migration models of prehistoric postmarital residence. American Journal of Physical Anthropology 77:471–482.

Konigsberg, L. W. 2000. Quantitative variation and genetics. *In* Human biology: An evolutionary and biocultural perspective. S. Stinson, B. Bogin, R. Huss-Ashmore, and D. O'Rourke, eds. Pp. 135–162. New York: Wiley-Liss.

Konigsberg, L. W., and S. D. Ousley. 1995. Multivariate quantitative genetics of anthropometric traits from the Boas data. Human Biology 67:481–498.

Korkhaus, G. 1930. Anthropologic and odontologic studies of twins. International Journal of Orthodontics 16:640–647.

Lanning, J. T. 1935. The Spanish missions of Georgia. Chapel Hill: University of North Carolina Press.

Larkford, G. E. III. 1984. Saying hello to the Timucua. Mid-America Folklore 12:7–23.

Larsen, C. S. 1982. The anthropology of St. Catherines Island: 3. Prehistoric biological adaptation. Anthropological Papers of the American Museum of Natural History, no. 57, pt 3.

Larsen, C. S. 1990. The archaeology of mission Santa Catalina de Guale: 2. Biocultural interpretations of a population in transition. Anthropological Papers of the American Museum of Natural History, no. 68.

Larsen, C. S. 1993. On the frontier of contact: Mission bioarchaeology in *La Florida*. *In* The Spanish missions of *La Florida*. B. G. McEwan, ed. Pp. 322–356. Gainesville: University Press of Florida.

Larsen, C. S. 2001. Bioarchaeology of Spanish Florida: The impact of colonialism. Gainesville: University Press of Florida.

Larsen, C. S., and D. E. Harn. 1994. Health in transition: Disease and nutrition in the Georgia Bight. *In* The diet and health of prehistoric Americans. K. D. Sobolik, ed. Pp. 222–234. Occasional paper no. 22. Carbondale: Center for Archaeological Investigations, Southern Illinois University–Carbondale.

Larsen, C. S., and D. L. Hutchinson. 1992. Dental evidence for physiological disruption: Biocultural interpretations from the eastern Spanish borderlands, U.S.A. Journal of Paleopathology 2:151–169.

Larsen, C. S., and G. R. Milner. 1994. In the wake of contact: Biological responses to conquest. New York: Wiley-Liss.

Larsen, C. S., and C. B. Ruff. 1994. The stresses of conquest in Spanish Florida: Structural adaptation and change before and after contact. *In* In the wake of contact: Biological responses to conquest. C. S. Larsen and G. R. Milner, eds. Pp. 21–34. New York: Wiley-Liss.

Larsen, C. S., and L. E. Sering. 2000. Inferring iron-deficiency anemia from human skeletal remains: The case of the Georgia Bight. *In* Bioarchaeological studies of life in the age of agriculture: A view from the Southeast. P. M. Lambert, ed. Pp. 116–133. Tuscaloosa: University of Alabama Press.

Larsen, C. S., and D. H. Thomas. 1982. The anthropology of St. Catherines Island: 4. The St. Catherines period mortuary complex. Anthropological Papers of the American Museum of Natural History. vol. 57, pt. 4.

Larsen, C. S., and D. H. Thomas. 1986. The archaeology of St. Catherines Island: 5. The South End Mound complex. Anthropological Papers of the American Museum of Natural History. no. 63, pt. 1.

Larsen, C. S., and T. A. Tung. 2002. Mission San Luis de Talimali: Final report on the human remains. Report submitted to the Florida Bureau of Archaeological Research, Florida Department of State, Tallahassee.

Larsen, C. S., with C. B. Ruff and M. C. Griffin. 1996. Implications of changing biomechanical and nutritional environments for activity and lifeway in the eastern Spanish borderlands. *In* Bioarchaeology of Native American adaptation in the Spanish borderlands. B. J. Baker and L. Kealhofer, eds. Pp. 95–125. Washington, DC: Smithsonian Institution Press.

Larsen, C. S., with R. Shavit and M. C. Griffin. 1991. Dental caries evidence for dietary change: An archaeological context. *In* Advances in dental anthropology. C. S. Larsen and M. Kelley, eds. Pp. 179–202. New York: Wiley-Liss.

Larsen, C. S., with C. B. Ruff, M. J. Schoeninger, and D. L. Hutchinson. 1992. Population decline and extinction in La Florida. *In* Disease and demography in the Americas. J. W. Verano and D. H. Ubelaker, eds. Pp. 25–39. Washington, DC: Smithsonian Institution Press.

Larsen, C. S., with A. W. Crosby, M. C. Griffin, D. L. Hutchinson, C. B. Ruff, K. F. Russell, M. J. Schoeninger, L. E. Sering, S. W. Simpson, J. L. Takacs, and M. F. Teaford. 2001. A biohistory of health and behavior in the Georgia Bight: The agricultural transition and the impact of European contact. *In* The backbone of history: Health and nutrition in the Western Hemisphere. R. H. Steckel and J. C. Rose, eds. Pp. 406–439. New York: Cambridge University Press.

Larson, L. H. 1969. Aboriginal subsistence technology on the Southeastern Coastal Plain during the late prehistoric period. Ph.D. dissertation, Department of Anthropology, University of Michigan.

Larson, L. H. 1978. Historic Guale Indians of the Georgia coast and the impact of the Spanish mission effort. *In* Tacachale: Essays on the Indians of Florida and southeastern Georgia during the historic period. J. Milanich and S. Proctor, eds. Pp. 120–140. Gainesville: University Presses of Florida.

Larson, L. H. 1980. Aboriginal subsistence technology on the southeastern coastal plain during the late prehistoric period. Gainesville: University Presses of Florida.

Laudonnière, R. 1975. Three voyages. Gainesville: University Presses of Florida.

Lewis, D. W., and R. M. Grainger. 1967. Sex-linked inheritance of tooth size. Archives of Oral Biology 12:539–544.
Lombardi, A. V. 1975. A factor analysis of morphogenetic fields in the human dentition. American Journal of Physical Anthropology 42:99–104.
Lundström, A. 1948. Tooth size and occlusion in twins. New York: S. Karger.
Lundström, A. 1967. Genetic aspects of variation in tooth width based on asymmetry and twin studies. Hereditas 57:403–410.
Lyon, E. 1976. The enterprise of Florida: Pedro Menéndez de Avilés and the Spanish Conquest of 1565–1568. Gainesville: University Presses of Florida.
Lyon, E. 1990. The enterprise of Florida. *In* Columbian consequences: 2. Archaeological and historical perspectives on the Spanish borderlands east. D. H. Thomas, ed. Pp. 281–296. Washington, DC: Smithsonian Institution Press.
Lyon, E. 1991. Pedro Menéndez de Avilés. *In* Spanish pathways in Florida: 1492–1992. A. L. Henderson and G. R. Mormino, eds. Pp. 94–117. Sarasota: Pineapple Press, Inc.
Magoon, D., with L. Norr, D. L. Hutchinson, and C. R. Ewen. 2001. An analysis of human skeletal materials from the Snow Beach Site (8Wa52). Southeastern Archaeology 20:18–30.
Marrinan, R. A. 1993. Archaeological investigations at mission Patale, 1984–1992. *In* The Spanish missions of *La Florida*. B. G. McEwan, ed. Pp. 244–294. Gainesville: University Press of Florida.
Marrinan, R. A., with J. A. Halpern, G. M. Heide, and C. Blackmore. 2000. Recent investigations at the O'Connell mission site (8Le157), Leon County, Florida. The Florida Anthropologist 53:224–249.
Mayhall, J. T. 1992. Techniques for the study of dental morphology. *In* Skeletal biology of past peoples: Research methods. S. R. Saunders and M. A. Katzenberg, eds. Pp. 59–78. New York: Wiley-Liss.
Mayhall, J. T., and L. Alvesalo. 1992. Dental morphology of 45,XO human females: Molar cusp area, volume, shape and linear measurements. Archives of Oral Biology 37:1039–1043.
Mayhall, J. T., with L. Alvesalo and G. Townsend. 1991. Tooth crown size in 46,Xi (Xq) human females. Archives of Oral Biology 36:411–414.
McEwan, B. G. 1991a. San Luis de Talimali: The archaeology of Spanish-Indian relations at a Florida mission. Historical Archaeology 25:36–60.
McEwan, B. G. 1991b. The archaeology of women in the Spanish New World. Historical Archaeology 25:33–41.
McEwan, B. G. 1992. Archaeology of the Apalachee village at San Luis de Talimali. Florida Archaeological Reports 28. Tallahassee: Florida Bureau of Archaeological Research.
McEwan, B. G. 1993. The Spanish missions of *La Florida*. Gainesville: University Press of Florida.
McEwan, B. G. 2000. The Apalachee Indians of Northwest Florida. *In* Indians of the greater Southeast. B. G. McEwan, ed. Pp. 57–84. Gainesville: University Press of Florida.
McEwan, B. G. 2001. The spiritual conquest of Florida. American Anthropologist 103: 633–644.
Mettler, L. W., with T. G. Gregg and H. E. Schaffer. 1988. Population genetics and evolution. Upper Saddle River, NJ: Prentice Hall.
Milanich, J. T. 1972. Excavations at the Richardson site, Alachua County, Florida: An early

17th century Potano Indian village (with notes on Potano culture change). Bureau of Historic Sites and Properties, Bulletin no. 2. Division of Archives, History, and Records Management, Tallahassee.

Milanich, J. T. 1978. The western Timucua: Patterns of acculturation and change. *In* Tacachale: Essays on the Indians of Florida and southeastern Georgia during the historic period. J. Milanich and S. Proctor, eds. Pp. 59–88. Gainesville: University Presses of Florida.

Milanich, J. T. 1990. The European entrada in La Florida: An overview. *In* Columbian consequences: 2. Archaeological and historical perspectives on the Spanish borderlands east. D. H. Thomas, ed. Pp. 3–30. Washington, DC: Smithsonian Institution Press.

Milanich, J. T. 1994. Archaeology of precolumbian Florida. Gainesville: University Press of Florida.

Milanich, J. T. 1996. The Timucua. Cambridge: Blackwell Publishers.

Milanich, J. T. 1999. Laboring in the fields of the Lord. Washington, DC: Smithsonian Institution Press.

Milanich, J. T., and C. Fairbanks. 1980. Florida archaeology. New York: Academic Press.

Milanich, J., and S. Proctor, eds. 1978. Tacachale: Essays on the Indians of Florida and southeastern Georgia during the historic period. Gainesville: University Presses of Florida.

Milanich, J. T., and W. C. Sturtevant. 1972. Francisco Pareja's 1613 confessionario: A documentary source for Timucuan ethnography. Tallahassee: Division of Archives, History, and Records Management, Florida Department of State.

Milner, G. R. 1980. Epidemic disease in the postcontact southeast: A reappraisal. Midcontinental Journal of Archaeology 5:39–56.

Milner, G. R. 1996. Prospects and problems in contact-era research. *In* Bioarchaeology of Native American adaptation in the Spanish borderlands. B. J. Baker and L. Kealhofer, eds. Pp. 198–208. Gainesville: University Press of Florida.

Moore, J. H. 1987. The Cheyenne nation: A social and demographic history. Lincoln: University of Nebraska Press.

Moore, J. H. 1994a. Ethnogenetic theory. National Geographic Research and Exploration 10:10–23.

Moore, J. H. 1994b. Putting anthropology back together again: The ethnogenetic critique of cladistic theory. American Anthropologist 96:925–948.

Moorrees, C. F. A. 1957. The Aleut dentition: A correlative study of dental characteristics in an Eskimoid people. Cambridge: Harvard University Press.

Moorrees, C. F. A., and R. B. Reed. 1964. Correlations among crown diameters of human teeth. Archives of Oral Biology 9:685–697.

Morner, M. 1967. Race mixture in the history of Latin America. Boston: Little Brown and Co.

Newton, S. D. 1937. On the mouths of triplets. Dental Record 57:364–368.

Niswander, J. D., and C. S. Chung. 1965. The effects of inbreeding on tooth size in Japanese children. American Journal of Human Genetics 17:390–398.

Oré. L. G. 1936. The martyrs of Florida. New York: Joseph F. Wagner.

Owen, R. C. 1965. The patrilocal band: A linguistically and culturally hybrid social unit. American Anthropologist 67:675–690.

Pearson, F. L. 1968. Spanish-Indian relations in Florida: A study of two visitas, 1657–1678. Ph.D. dissertation, University of Alabama.

Petersen, H. C. 2000. On statistical methods for comparison of intrasample morphometric variability: Zalavár revisted. American Journal of Physical Anthropology 113:79–84.

Potter, R. H., and W. E. Nance. 1976. A twin study of dental dimension I. Discordance, asymmetry, and mirror imagery. American Journal of Physical Anthropology 44:391–396.
Potter, R. H., with W. E. Nance, P. L. Yu, and W. B. Davis. 1976. A twin study of dental dimension II. Independent genetic determination. American Journal of Physical Anthropology 44:397–412.
Potter, R. H., with J. P. Rice, A. A. Dahlberg, and T. Dahlberg. 1983. Dental size traits within families: Path analysis for first molar and lateral incisor. American Journal of Physical Anthropology 61:283–289.
Potter, R. H., with P. L. Yu, A. A. Dahlberg, A.D. Merritt, and P. M. Conneally. 1968. Genetic studies of tooth size factors in Pima Indian families. American Journal of Human Genetics 20:89–99.
Quinn, W. W. 1993. Intertribal integration: The ethnological argument in Duro v. Reina. Ethnohistory 40:34–69.
Ramenofsky, A. F. 1987. Vectors of death: The archaeology of European contact. Albuquerque: University of New Mexico Press.
Ramenofsky, A. F., with A. K. Wilbur and A. C. Stone. 2003. Native American disease history: Past, present, and future directions. World Archaeology 35:241–257.
Rebich, T., and M. Markovic. 1976. Inheritance of tooth dimension: A quantitative genetic twin study. 52nd Congress, June 28–July 2, Transactions of the European Orthodontic Society 59–68.
Reitz, E. J. 1993. Evidence for animal use at the missions of Spanish Florida. *In* The Spanish Missions of *La Florida*. B. G. McEwan, ed. Pp. 376–398. Gainesville: University Press of Florida.
Relethford, J. H. 1980. Bioassay of kinship from continuous traits. Human Biology 52: 689–700.
Relethford, J. H. 1996. Genetic drift can obscure population history: Problem and solution. Human Biology 68:29–44.
Relethford, J. H. 2003. Anthropometric data and population history. *In* Human biologists in the archives: Demography, health, nutrition, and genetics in historical populations. D. A. Herring and A. C. Swedlund, eds. Pp. 32–52. Cambridge: Cambridge University Press.
Relethford, J. H. 2004. Boas and beyond: Migration and craniometric variation. American Journal of Human Biology 16:379–386.
Relethford, J. H., and J. Blangero. 1990. Detection of differential gene flow from patterns of quantitative variation. Human Biology 62:5–25.
Ruff, C. B., and C. S. Larsen. 1990. Postcranial biomechanical adaptations to subsistence strategy changes on the Georgia coast. *In* The archaeology of mission Santa Catalina de Guale: 2. Biocultural interpretations of a population in transition. C. S. Larsen, ed. Pp. 36–49. New York: Anthropological Papers of the American Museum of Natural History, no. 68.
Ruff, C. B., and C. S. Larsen. 2001. Reconstructing behavior in Spanish Florida: The biomechanical evidence. *In* Bioarchaeology of Spanish Florida: The impact of colonialism. C. S. Larsen, ed. Pp. 113–145. Gainesville: University Press of Florida.
Ruff, C. B., with C. S. Larsen and W. C. Hayes. 1984. Structural changes in the femur with the transition to agriculture on the Georgia coast. American Journal of Physical Anthropology 64:125–136.
Ruhl, D. L. 2000. Recent archaeobotanical findings at mission San Luis de Apalachee. Manuscript on file, Florida Bureau of Archaeological Research, Tallahassee.

Russell, K. F. 1987. Demographic implications of European contact in northern Spanish Florida. American Journal of Physical Anthropology 72:248.

Russell, K. F., with I. Choi and C. S. Larsen. 1990. The paleodemography of Santa Catalina de Guale. *In* The archaeology of mission Santa Catalina de Guale: 2. Biocultural interpretations of a population in transition. C. S. Larsen, ed. Pp. 94–120. New York: Anthropological Papers of the American Museum of Natural History, no. 68.

Sattenspiel, L., and H. Harpending. 1983. Stable populations and skeletal age. American Antiquity 48:489–498.

Sattler, R. A. 1992. Ethnic transformation on the Gulf Coast: The Apalachi case. Paper presented at the Annual Meeting of the American Association for the Advancement of Science, Chicago.

Saunders, R. 1992. Continuity and change in Guale Indian pottery, A.D. 1350–1702. Ph.D. dissertation, Department of Anthropology, University of Florida.

Saunders, R. 1993. Architecture of the missions Santa María and Santa Catalina de Amelia. *In* The Spanish missions of *La Florida.* B. G. McEwan, ed. Pp. 35–61. Gainesville: University Press of Florida.

Saunders, R. 2000. Stability and change in Guale Indian pottery A.D. 1300–1702. Tuscaloosa: University of Alabama Press.

Scarry, C. M. 1993. Plant production and procurement in Apalachee Province. *In* The Spanish missions of *La Florida.* B. G. McEwan, ed. Pp. 357–375. Gainesville: University Press of Florida.

Scarry, J. F. 1992. Political offices and political structure: Ethnohistoric and archaeological perspectives on the native lords of Apalachee. *In* Lords of the Southeast: Social inequality and the Native elites of southeastern North America. A. W. Barker and T. R. Pauketat, eds. Pp. 163–183. Archaeological Papers of the American Anthropological Association no. 3.

Schmidt, C. W. 1993. Paleodemographic analysis of the Santa Catalina de Guale de Santa Maria ossuary. M.A. thesis, Department of Anthropology, Purdue University.

Schnutenhaus, S., and F. W. Rösing. 1998. World variation of tooth size. *In* Dental anthropology: Fundamentals, limits, and prospects. K. W. Alt, F. W. Rösing, and M. Teschler-Nicola, eds. Pp. 521–535. New York: Springer.

Schoeninger, M. J., with N.J. van Der Merwe, K. Moore, J. Lee-Thorp, and C. S. Larsen. 1990. Decrease in diet quality between the prehistoric and the contact periods. *In* The archaeology of mission Santa Catalina de Guale: 2. Biocultural interpretations of a population in transition. C. S. Larsen, ed. Pp. 78–93. New York: Anthropological Papers of the American Museum of Natural History, no. 68.

Schultz, M., with C. S. Larsen and K. Kreutz. 2001. Disease in Spanish Florida: Microscopy of porotic hyperostosis and cribra orbitalia. *In* Bioarchaeology of Spanish Florida: The impact of colonialism. C. S. Larsen, ed. Pp. 207–225. Gainesville: University Press of Florida.

Scott, G. R., and G. C. Turner. 1997. The anthropology of modern human teeth. Cambridge: Cambridge University Press.

Shapiro, G. 1987. Archaeology at San Luis: Broad-scale testing, 1984–1985. Florida Archaeology no. 3. Tallahassee: Florida Bureau of Archaeological Research.

Shapiro, G., and R. Vernon. 1992. The church complex. Florida Archaeology no. 6. Tallahassee: Florida Bureau of Archaeological Research.

Sharma, K., with R. S. Corruccini and A. M. Henderson. 1985. Genetic variance in dental dimensions of Punjabi twins. Journal of Dental Research 64:1389–1391.

Sharrock, S. R. 1974. Crees, Cree-Assiniboines, and Assiniboines: Interethnic social organization on the far northern plains. Ethnohistory 21:95–122.

Shepard, H. E. 2003. Geometry in Apalachee buildings at mission San Luis. Southeastern Archaeology 22:165–176.

Siebert, W. H. 1940. The departure of the Spaniards and other groups from east Florida, 1763–1764. Florida Historical Quarterly 19:145–154.

Simpson, S. W. 2001. Patterns of growth disruption in La Florida: Evidence from enamel microstructure. *In* Bioarchaeology of Spanish Florida: The impact of colonialism. C. S. Larsen, ed. Pp. 146–180. Gainesville: University Press of Florida.

Simpson, S. W., with D. L. Hutchinson and C. S. Larsen. 1990. Coping with stress: Tooth size, dental defects and age-at-death. *In* The archaeology of Mission Santa Catalina de Guale: 2. Biocultural interpretations of a population in transition. C. S. Larsen, ed. Pp. 66–77. Anthropological Papers of the Museum of Natural History, no. 68.

Sjøvold, T. 1984. A report on the heritability of some cranial measurements and non-metric traits. *In* Multivariate statistical methods in physical anthropology. G. N. van Vark and W. W. Howells, eds. Pp. 233–246. Dordrecht: D. Reidel Publishing Company.

Sluiter, E. 1985. The Florida situado: Quantifying the first eighty years, 1571–1651. Research Publications of the P. K. Yonge Library of Florida History. Gainesville: University of Florida.

Smith, H. G., and M. Gottlob. 1978. Spanish-Indian relationships: Synoptic history and archeological evidence, 1500–1763. *In* Tacachale: Essays on the Indians of Florida and southeastern Georgia during the historic period. J. Milanich and S. Proctor, eds. Pp. 1–18. Gainesville: University Presses of Florida.

Smith, M. T. 1987. Archaeology of aboriginal culture change in the interior southeast: Depopulation during the early historic period. Gainesville: University Presses of Florida.

Snow, D. R., and K. M. Lanphear. 1988. European contact and Indian depopulation in the northeast: The timing of the first epidemics. Ethnohistory 35:15–33.

Snow, D. R., and K. M. Lanphear. 1989. "More methodological perspectives": A rejoinder to Dobyns. Ethnohistory 36:299–304.

Sokal, R. R., and F. J. Rohlf. 1995. Biometry. New York: W. H. Freeman and Company.

Sparks, C. S., and R. L. Jantz. 2002. A reassessment of human cranial plasticity: Boas revisited. Proceedings of the National Academy of Sciences USA 99:14636–14639.

Spellman, C. W. 1965. The "Golden Age" of the Florida mission, 1632–1674. Catholic Historical Review 51:355–378.

Stannard, D. E. 1991. The consequences of contact: Toward an interdisciplinary theory of native responses to biological and cultural invasion. *In* Columbian consequences: 3. The Spanish borderlands in Pan-American perspective. D. H. Thomas, ed. Pp. 519–560. Washington, DC: Smithsonian Institution Press.

Steadman, D. W. 2001. Mississippians in motion? A population genetic analysis of interregional gene flow in west-central Illinois. American Journal of Physical Anthropology 114:61–73.

Steinen, K. T., and R. Ritson. 1996. In defense of the frontier: Considerations of Apalache warfare during the period 1539–1540. Florida Anthropologist 49:111–120.

Stojanowski, C. M. 2001. Cemetery structure, population aggregation and biological vari-

ability in the mission centers of *La Florida.* Ph.D. dissertation, Department of Anthropology, University of New Mexico.
Stojanowski, C. M. 2003. Matrix decomposition model for investigating prehistoric intracemetery biological variation. American Journal of Physical Anthropology 122:216–231.
Stojanowski, C. M. 2004. Population history of native groups in pre- and postcontact Spanish Florida: Aggregation, gene flow, and genetic drift on the southeastern U.S. Atlantic Coast. American Journal of Physical Anthropology 123:316–322.
Stojanowski, C. M., with C. S. Larsen, T. A. Tung, and B. G. McEwan. n.d. Biological affinity of the coffin burials at mission San Luis: Hispanic residents or native elites? Ms. on file, Department of Anthropology, Southern Illinois University.
Storey, R. 1986. Diet and health comparisons between pre- and post-Columbian Americans in North America. American Journal of Physical Anthropology 69:268.
Sturtevant, W. C. 1962. Spanish-Indian relations in southeastern North America. Ethnohistory 9:41–94.
Susanne, C. 1977. Heritability of anthropological characters. Human Biology 49:573–580.
Swadesh, M. 1964. Linguistic overview. *In* Prehistoric man in the New World. J. D. Jennings and E. Norbeck, eds. Pp. 527–556. Chicago: University of Chicago Press.
Swanton, J. R. 1922. Early history of the Creek Indians and their neighbors. Washington, DC: Smithsonian Institution Press.
Swanton, J. R. 1929. The Tawasa language. American Anthropologist 31:435–453.
Swanton, J. R. 1946. The Indians of the Southeastern United States. Washington, DC: Smithsonian Institution Press.
Teaford, M. F., with C. S. Larsen, R. F. Pastor, and V. E. Noble. 2001. Pits and scratches: microscopic evidence of tooth use and masticatory behavior in La Florida. *In* Bioarchaeology of Spanish Florida: The impact of colonialism. C. S. Larsen, ed. Pp. 82–112. Gainesville: University Press of Florida.
Terrell, J. E., with T. L. Hunt, and C. Gosden. 1997. The dimensions of social life in the Pacific. Current Anthropology 38:155–195.
Thomas, D. H. 1988. St. Catherines: An island in time. Georgia History and Culture Series. Atlanta: Georgia Endowment for the Humanities.
Thomas, D. H. 1990a. The Spanish missions of La Florida: An overview. *In* Columbian consequences: 2. Archaeological and historical perspectives on the Spanish borderlands east. D. H. Thomas, ed. Pp. 357–398. Washington, DC: Smithsonian Institution Press.
Thomas, D. H. 1990b. Introduction. *In* The archaeology of mission Santa Catalina de Guale: 2. Biocultural interpretations of a population in transition. C. S. Larsen, ed. Pp. 8–10. Anthropological Papers of the American Museum of Natural History, no. 68.
Townsend, G., and L. Alvesalo. 1985a. Tooth size in 47,XYY males—evidence for a direct effect of the Y chromosome on growth. Australian Dental Journal 30:268–272.
Townsend, G., and L. Alvesalo. 1985b. The size of permanent teeth in Klinefelter (47,XXY) syndrome in man. Archives of Oral Biology 30:83–84.
Townsend, G. C., and T. Brown. 1978a. Heritability of permanent tooth size. American Journal of Physical Anthropology 49:497–504.
Townsend, G. C., and T. Brown. 1978b. Inheritance of tooth size in Australian Aborigines. American Journal of Physical Anthropology 48:305–314.
Townsend, G. C., and T. Brown. 1979. Family studies of tooth size factors in the permanent dentition. American Journal of Physical Anthropology 50:183–190.

Townsend, G. C., with B. L. Jensen and L. Alvesalo. 1984. Reduced tooth size in 45,X (Turner syndrome) females. American Journal of Physical Anthropology 65:367–372.
Townsend, G. C., with L. Richards and T. Hughes. 2003. Molar intercuspal dimensions: genetic input to phenotypic variation. Journal of Dental Research 82:350–355.
Townsend, G. C., T. Brown, L. C. Richards, J. R. Rogers, S. K. Pinkerton, G. R. Travan, and V. B. Burgess. 1986. Metric analyses of the teeth and faces of South Australian twins. Acta Geneticae Medicae et Gemellologiae 35:179–192.
Ubelaker, D. H. 1978. Human skeletal remains. Washington, DC: Taraxacum.
Van Reenen, J. F. 1961. The use of the concept of tooth material as an indication of tooth size in a group of Kalahari Bushmen. South African Journal of Science 57:347–352.
Van Reenen, J. F. 1964. Dentition, jaws and palate of the Kalahari Bushman. Journal of the Dental Association of South Africa 19:38–44.
Varner, J., and J. G. Varner. 1951. Florida of the Inca. Austin: University of Texas Press.
Varrela, J., with G. Townsend and L. Alvesalo. 1988. Tooth crown size in human females with 45,X/46,XX chromosomes. Archives of Oral Biology 33:291–294.
Walker, P. L. 2001. A Spanish Borderlands perspective on La Florida bioarchaeology. *In* Bioarchaeology of Spanish Florida: The impact of colonialism. C. S. Larsen, ed. Pp. 274–307. Gainesville: University Press of Florida.
Weddle, R. S. 1985. Spanish sea: The Gulf of Mexico in North American discovery, 1500–1685. College Station: Texas A&M University Press.
Weisman, B. R. 1992. Excavations on the Franciscan frontier: Archaeology at the Fig Springs mission. Gainesville: University Press of Florida.
Wenhold, L. L. 1936. A 17th century letter of Gabriel Díaz Vara Calderón, bishop of Cuba, describing the Indians and Indian missions of Florida. Smithsonian Miscellaneous Collection, vol. 95, no. 16. Washington, DC: Smithsonian Institution Press.
Wilkinson, L., with G. Blank and C. Gruber. 1996. Desktop data analysis with SYSTAT. New York: Prentice Hall.
Wobst, H. M. 1974. Boundary conditions for Paleolithic social systems: A simulation approach. American Antiquity 39:147–178.
Wood, B. F., and L. J. Green. 1969. Second premolar morphologic trait similarities in twins. Journal of Dental Research 48:74–78.
Wood, J. W., with G. R. Milner, H. C. Harpending, and K. M. Weiss. 1992. The osteological paradox. Current Anthropology 33:343–370.
Worth, J. E. 1992. The Timucuan missions of Spanish Florida and the rebellion of 1656. Ph.D. dissertation, Department of Anthropology, University of Florida.
Worth, J. E. 1995. The struggle for the Georgia coast: An eighteenth-century Spanish retrospective on Guale and Mocama. American Museum of Natural History, Anthropological papers, no. 75.
Worth, J. E. 1998a. The Timucuan chiefdoms of Spanish Florida: 1. Assimilation. Gainesville: University Press of Florida.
Worth, J. E. 1998b. The Timucuan chiefdoms of Spanish Florida: 2. Resistance and destruction. Gainesville: University Press of Florida.
Worth, J. E. 2001. The ethnohistorical context of bioarchaeology in Spanish Florida. *In* Bioarchaeology of Spanish Florida: The impact of colonialism. C. S. Larsen, ed. Pp. 1–21. Gainesville: University Press of Florida.
Wright, J. Leitch. 1981. The only land they knew. New York: The Free Press.

Zilberman, U., with N. Abramovitz and E. Mass. 1995. Cusp relations in familial dysautonomy compared to normal population. *In* Proceedings of the 10th international symposium on dental morphology. R. J. Radlanski and H. Renz, eds. Pp. 136–139. Berlin.

Zubillaga, F. 1946. Monumenta antiquae Florida (1566–1572). Monumenta Historica Societatis Iesu, vol. 69. Monumenta Missionum Societatis Iesu vol. 3, Rome.

Index